# Why Brains
# Need Friends

# Why Brains Need Friends

## The Neuroscience of Social Connection

# Ben Rein, PhD

Avery
an imprint of Penguin Random House
New York

AVERY

an imprint of Penguin Random House LLC
1745 Broadway, New York, NY 10019
penguinrandomhouse.com

*Book design by Angie Boutin*

Library of Congress Cataloging-in-Publication Data

Names: Rein, Ben, author.
Title: Why brains need friends: the neuroscience of social connection / Ben Rein, PhD.
Description: New York: Avery, [2025] | Includes index.
Identifiers: LCCN 2025002612 (print) | LCCN 2025002613 (ebook) | ISBN 9780593850848 hardcover | ISBN 9780593850855 ebook
Subjects: LCSH: Social interaction—Psychological aspects | Social interaction—Health aspects | Brain | Well-being
Classification: LCC HM1111 .R45 2025 (print) | LCC HM1111 (ebook) | DDC 302—dc23/eng/20250627
LC record available at https://lccn.loc.gov/2025002612
LC ebook record available at https://lccn.loc.gov/2025002613

Printed in the United States of America
1st Printing

The authorized representative in the EU for product safety and compliance is Penguin Random House Ireland, Morrison Chambers, 32 Nassau Street, Dublin D02 YH68, Ireland, https://eu-contact.penguin.ie.

*For those who have struggled with loneliness,*
*and for everyone who showed them the way out.*

# Contents

# I
## How Interaction and Isolation Shape Brain Health

# II
## Nurturing Your Brain in a Post-Interaction World

# "NO BIG WORDS" CLAUSE

———— ○ ————

It always bothers me when people use big words for no reason.

Simpler terms almost always get the job done better. Why "utilize" or "employ" big words when we can just *use* smaller ones? Sure, it might feel good to "exhibit" that you "wield" a "capacious" vocabulary, but you can still be a good writer with basic language. Big words tend to "obfuscate"—I mean *cloud*—the meaning of ideas. Sometimes they help make a point, but they can also box out readers.

Something similar happens in science. Scientific jargon has blurred the meaning of academic papers, preventing the public from following along. We must be better at communicating. Instead of assuming readers know what a presynaptic vesicle is, we can simply call it "a little pouch of neurotransmitters" and be on our merry way. Finding the simple way to explain things is far more constructive than packing sentences with jargon and hoping for the best. If scientists want to make a difference in the world, we must leave the gates open. In this book, you're all invited in. Don't get me wrong, we'll cover tons of serious neuroscience, but it will be written for all audiences to appreciate. Just as doctors take the Hippocratic oath to *do no harm,* I am taking the oath to *use no jargon.*

With that said, some jargon is inevitable. Certain terms just *need* to be used. There's no other way to say GABA (short for gamma-aminobutyric acid) than by saying GABA. But terminology can always be *explained.* As I wrote this book, I found myself using some scientific terms, so to make sure that nobody is left behind, I've written an appendix, which can be found on my website (benrein.com/book). In the appendix, I explain what several terms really mean and sometimes provide added topical commentary. If you see a plus sign on a word like this,[+] it means there's something in the appendix about it. I hope to see you there.

# INTRODUCTION

—————— ○ ——————

## Three Hard Truths About Our Social Lives

### "HELLO?"

Picture this: It's four p.m. on a Thursday, and you're hunched over a computer feeling stressed and anxious. This week has been a nightmare; you're working overtime on a huge project that totally *sucks*, and the deadline is coming up fast. With your stress mounting, you find it harder and harder to focus. You can't wait for this project to be over, but it's hard to imagine how you'll get through it.

Feeling depleted, you lean back in your chair and close your eyes, taking a breath. Your body is begging you for rest, but you convince yourself that if you just work for a few more hours tonight, you'll feel better tomorrow. You lean forward, refocusing your weary eyes on the screen. It feels wrong to go against your gut, but do you really have another choice?

As you reach this decision, your phone starts to buzz and light up on your desk. To your surprise, it's an incoming call from an old friend you haven't spoken to in years. What the heck could this be about? You glance back and forth between your phone and your computer, juggling the decision. Do you keep banging away at the project or take the call? For a moment, really try to put yourself in this position, and be honest: **Would you pick up the phone?**

It's a tricky situation. On the one hand, you desperately want to get through this project. A distraction could be catastrophic. On the other hand, you're curious why they're calling. Maybe it's good news? Or . . . what if it's bad news? Are they just going to ask a favor? What if the conversation is awkward and you have to wiggle your way out?

Despite these concerns, something tells you to pick up the phone. The impulse takes over, and you answer the call.

"Hello?"

"Hey!"

What follows is a surprisingly fruitful and satisfying conversation. There's no favor to be asked, no bad news to be shared, and no awkwardness to the exchange. Your old friend was simply thinking of you and wanted to check in. Together, you reminisce on familiar memories, laughing and sharing stories from the past. Before long, you have turned away from your computer and are engaged fully in the call, pouring your attention into the voice on the other end. After nearly an hour, you say your goodbyes and promise to call again soon. You really mean it.

Returning to your computer, you feel unexpectedly renewed and energized. Your mind is clear and you're ready to take on the project, which suddenly doesn't seem so overwhelming or daunting. You work comfortably for another hour or two, making far more progress than you'd expected.

◯◯

We've all experienced something like this, right? Sometimes a good conversation is the perfect remedy for a sour mood, especially after you've gone a few days without a satisfying interaction. This is more than just a relatable anecdote; it's supported by science. Studies show that people tend to leave conversations in a better mood and with less stress. These effects can compound over the long term: People who have more frequent interactions report greater well-being, while those with unmet social needs score lower.

Yes, interactions are natural mood-boosters, but that's not all they can do for us. People who interact more are also at lower risk for dementia, heart failure, diabetes, depression, and anxiety. The support we get from our social systems has been shown to reduce vulnerability to stress and to increase pain tolerance. On the other hand, social *isolation* is one of the strongest known predictors of suicide. Clearly our social lives have a strong hand in shaping our health and well-being.

But that's not all. Research suggests that our social lives might also shape our *lifespan*. It may sound ludicrous, but whether you tend to answer that call or let it buzz to your voicemail could influence the amount of time you have on Earth. One study tracked over three hundred thousand people for 7.5 years on average, during which some of the subjects naturally passed away. Remarkably, it was found that people who had weaker social relationships were *50 percent more likely to die* during the study. To put that into perspective, that makes being isolated roughly *twice* as bad for you as being obese, and *four* times as bad as living in a highly polluted area.

Now, rewinding to our phone-call dilemma, were you thinking about your health while deciding whether to answer? Probably not, which would be entirely normal. We just don't think about our social lives like this . . . but we *should*. As a neuroscientist studying the biology of social behavior, I believe that interpersonal connections are just as important as other pillars of health like exercise, sleep, and nutrition. However, we fail to prioritize them as such.

This is precisely the challenge we face today. Our world is currently facing a social problem, and the sooner we acknowledge that reality, the sooner we can begin working against it. This brings me to **Hard Truth No. 1: We live in a divided world.** There are many scapegoats we can point fingers at: an ever-growing surge in social media use, the COVID-19 pandemic, the rise of remote work, political polarization, and countless others. Regardless of who or what is to blame, we must recognize and appreciate that our social lives are receding, and that's a huge problem.

Luckily, I do think this recognition is happening. I've been relieved to see the public conversation around isolation advance over the last few years. We've started to notice the headlines and hear the podcasts about our social problem. The best part is, there is an extremely obvious and simple solution: *Socialize more!* But have we really acted on this message? Has anything changed? Personally, I've seen very little in the way of community-building and coming together, only the continued presence of division forking our society. Why aren't we doing enough to fix this? One guess is that we're not fully appreciating the consequences of *not* interacting, so I'm here to bear some sour news that you probably don't want to hear. *We must take this issue seriously.* That's because of **Hard Truth No. 2: Division is the enemy of brain health.**

The human brain has been shaped through evolution to *reward* us for connection and *punish* us for isolation. As a result, we have so much to gain from socializing, and arguably even more to lose without it. We all know that we should really answer the phone *more* and make time for friends and family, but we're seemingly not motivated enough to do something about it. I believe that part of the reason we've failed to act is because the current discussion around isolation is incomplete. We've heard that interaction is good for us and isolation is bad, but what does that really look like in the brain and body? The articles and podcasts haven't really incorporated the neuroscience that shows us what's truly at stake, and there's a lot to share. Social connection penetrates deep into our inner workings, touching on countless systems within us. When we fail to prioritize connection, our biology can struggle in ways most people don't realize. I believe that if we truly appreciate what's at risk, we may be more motivated to act. After all, how can you be motivated to do something if you don't know why you're doing it?

Now, there's one more hard truth to accept, and it will be the most difficult pill to swallow. While we tend to blame *external factors* like COVID-19 and remote work for our social problem, we must also recognize our own role. Modern humans behave in ways that can cast us into division, but it's not all our fault. The human brain was shaped in a very different world from the one we inhabit today, and because of that, it doesn't always do what we would consider the *right* or the *best* thing in today's society. For example, we sometimes argue with strangers on the internet, we tend to empathize less with people who are *unlike* us, and we underestimate the value of giving compliments. These social pitfalls are obviously unhelpful for building connection, but they're simply the result of how the brain is wired. As smart and capable as the brain is, it's not a perfect organ. It has flaws that can get in the way of our connections. This brings up **Hard Truth No. 3: The brain has internal shortcomings that can drive us apart.**

If we hope to construct a society in which we truly choose and prioritize connection, I believe we must identify these barriers and address them head-on. Throughout this book, we will outline numerous bizarre things the brain does that can lead to division, explore why these pitfalls are built into our minds, and discuss what we can do about them. My goal is that by shining our headlights on these potholes in our social brains, we might be able to swerve and avoid them through thoughtful and deliberate action.

In today's world, every person on Earth shares a common enemy: **division**. Humanity is quietly at war with an opponent that threatens our brain health and the future of our species. There's only one way to defeat this opponent, and it's incredibly simple: by coming together. This book will pull back the curtain on mountains of neuroscience research to help you understand the true nature of this battle. Together, we will ask: *What*, on a biological level, do interactions offer for the brain? What happens when we spend time with others, or when we're left alone for too long? Does interacting with people *online* do anything for the brain? Does hanging out with your dog count? And what forces are lurking insidiously beneath the surface, tinkering with our brains to sabotage our connections? At times you will be shocked and mind-blown by the amazing science we'll cover. You'll come to question things you used to believe, and I hope you'll find wonder in the incredible systems that operate within us.

Most important of all, if I do my job well enough, I hope you will be motivated to join hands with your fellow humans and join the fight against our common enemy.

⟲

I've been fascinated by social interactions since I was a child. I distinctly remember looking around my elementary school cafeteria and being captivated by the many tables of kids eating lunch around me. Some tables were quiet, occupied by the timidest of students, who sat in groups only because it was the appropriate thing to do. They would read books or stare down at their sandwiches to avoid conversation, and when the scarce interactions did occur, they were clumsy and unpleasant. Being one of the kids who read a lot of books, I often found myself seated at these tables. However, I felt socially deprived. I spent my lunch periods looking around and imagining myself sitting with my more animated and popular classmates. Everything was louder and more fun at those tables. It was a completely different social environment . . . one that I desperately wanted to be a part of.

Over several years of people-watching in childhood, I recognized that people have all sorts of different social habits. I started to imagine that *sociability exists on a continuum*—ranging from the quiet depths of shyness to the extreme outer limits of extroversion. Everyone exists somewhere on this spectrum. Where are you?

Introverted                                        Extroverted

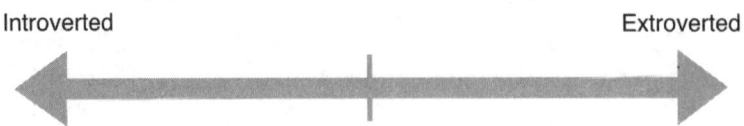

I found myself returning to this idea over and over as I grew older, wanting to understand *why* we're all so different. Fortunately I did eventually find my way to tables with other extroverts, but I never stopped looking around. I never stopped observing the beautiful variability of social behaviors on display, wondering about the operating systems that make us each unique.

When I got to college, I chose psychology as my major in hopes that maybe one day I could study those social differences. However, something felt off about my education. It felt sort of *incomplete*, like I was studying the wrong thing. It didn't excite me in the way I expected. As I progressed through my courses, I slowly realized what it was: I was learning about *behavior* and why people did things, but I was truly more interested in the *brain*. Thinking back to my childhood idea of a *social continuum*, I realized that each student's unique way of expressing themselves reflected some invisible differences in how their *brain* was functioning! I became fixated on pinpointing those differences. Why do some students' brains feel more comfortable at the quiet tables, while others prefer to be interactive? Can we trace these social habits back to specific brain systems? It seemed that I had discovered my true passion in studying the brain, but there was a big problem . . . neuroscience scared the shit out of me.

I was completely intimidated by the subject. Changing my major to neuroscience would mean taking courses like biochemistry, genetics, and molecular biology, and I simply didn't think I had what it took. Having struggled with self-confidence through my early life, I didn't believe in myself. So instead of pursuing my dream, I continued studying psychology, hoping it would work out in the end. As much as I feared I was making a mistake, it felt like my only choice. I had a good GPA, and I didn't want to risk ruining it by switching to neuroscience. With my head down, I sauntered toward a future I wasn't sure I wanted.

Then, everything changed. Just three semesters before graduation, I experienced a terrifying nightmare that completely altered my life. In the dream I was an adult, with a family and a home. Everything was superficially normal, except I was haunted by an evil force that I couldn't see, but it controlled my life. If I disobeyed this force, I would begin to shapeshift in horrifying ways: My arms and legs would bend in the wrong directions, my face would swell with fluid until the skin neared its breaking point, and sores would appear all over my skin. It was just as horrible as it sounds. At a pivotal point in the dream, the demonic force summoned me into the basement of my home. While it had always been invisible to me, I suddenly understood that it was about to reveal itself. I felt an incredible, terrible power surrounding me, filling me with dread and awe. My fear was indescribable. Then just as I felt the force overtake me, I awoke.

The sensation I experienced upon waking was equally powerful. I shot up from my creaky twin bed into a silent dark room. It was the middle of the night, but my body was electrified with adrenaline. I lay back down and attempted to gather myself, but I couldn't shake the sensation of the nightmare. For some reason—of all the thoughts I could have had when I woke up—the one thing I couldn't stop thinking about was, *How the hell did my brain do that?* I was amazed that a single organ could both *generate* such a complex nightmare—complete with vivid landscapes, unpredictable characters, and a compelling storyline—and simultaneously *experience* this dream, interacting with the characters and making decisions. How was all of this happening inside my brain?

That curiosity was the last straw. I was officially pushed over the edge. It was clear that I could no longer suppress my true desire to study neuroscience. I stayed up all night planning how I would change my academic path. The very next day, I spoke to my academic advisor, joined the student neuroscience club, and started reaching out to neuroscience labs to volunteer. I planned to continue studying social behavior, but now through a *neuroscience* lens. I hoped that maybe, just maybe, I would one day figure out why some students' brains prefer to stare down at their sandwiches, while others' like to play rowdily.

Looking back on my nightmare, I always wondered if that evil "force" was meant to represent my career. Perhaps my subconscious mind intended to warn me that I was heading toward an uncomfortable future, where my career dominated my life. Being in the wrong career, I would have to do things I didn't want to do and be transformed into a version of myself I don't identify with. Maybe it was never going to reveal itself to me in the dream after all. Instead, perhaps it was planning to reveal the truth to me in real life, in those moments after I woke up and couldn't stop thinking about the brain. I also wonder if perhaps the force I nearly met in the dream somehow *was* my subconscious, coming to scare me into noticing the mistake I was making.

Of course, this story has a happy ending. I went on to do a PhD in neuroscience at SUNY Buffalo, and I did just fine in those scary courses with big long names that I thought I'd fail. After my PhD, I was hired at Stanford University to continue working as a neuroscientist. I'm honored to say that I've published over twenty scientific articles examining how the social brain works. My research has explored how our genes shape our social behaviors, how drugs like MDMA enhance empathy, how social motivation can be influenced by environmental pressures, why people treat one another brutally on social media, how interpersonal interventions can prevent suicides, and even how to measure social interactions in mice. I'm proud of my younger self for having the courage to take on a challenge that felt terrifying, but necessary.

Through my service in the field of neuroscience, I've learned amazing things about the science of our interactions that I can't wait to share with you in this book. And if you're anything like me in that you find neuroscience intimidating, don't worry. This book is for you. Aside from my work in the lab, I've also had the delight of teaching neuroscience on social media, sharing the beauty and mystery of the brain with anyone willing to listen. I've challenged myself to develop my science communication skills because I believe that anyone can learn about the brain if the information is presented the right way. We all have one, and we deserve to understand it. I've written this book accordingly, so that everyone can access that knowledge.

It's not breaking news that we have a social problem. I'm sure many books published in the coming years will diagnose just how isolated we've become. This is not one of them. This book isn't just here to tell you things you already know, about how lonely we are and how the modern world is changing. I intend to dive deeper, guiding you through the mysterious biology of the brain and explaining the significance of these changes for our health and well-being. These pages will not merely explain that you need connection but will show you *why*. We will answer myriad questions about the pink squishy machine in your head that desperately wants company. We'll cover a tremendous amount of science carried out over several decades across multiple continents. And while you probably haven't heard of these studies before or read them yourself, they hold the potential to seriously change your life.

## Key Takeaways

1. Social connection is crucial for well-being. It's not just a nice-to-have; it's as vital as exercise, sleep, and nutrition. However, we tend to neglect it as a key component of our health.

2. Social connection has profound health benefits. It's linked to a lower risk of dementia, heart failure, diabetes, depression, and anxiety. It is even associated with increased pain tolerance and a longer lifespan.

3. The public conversation around isolation has failed to incorporate the underlying neuroscience. This book will clarify what's truly at stake.

4. We must recognize three hard truths about social interaction:

   **1: We live in a divided world.** External factors like social media, the pandemic, and political polarization have widened our social gaps.

   **2: Division is the enemy of brain health.** The brain is wired to reward connection and punish isolation, making division a serious threat to well-being.

   **3: The brain has internal shortcomings that can drive us apart.** Our neural wiring has bugs and biases that can hinder connection. Understanding these natural shortcomings can help us build more effective social lives.

To view the references cited in this chapter, please visit benrein .com/book.

# I

## How Interaction
## and Isolation Shape
## Brain Health

# DON'T BE SHY

The Hidden Benefits of Social Interaction

## A POST-INTERACTION WORLD

Your brain is a prediction machine. While it sits silently behind your eyes, it works tirelessly to gather information from the world around you and guess what might happen next. This is good: It lets you make faster decisions and adapt to changing environments. Just think of Tom Hanks in the movie *Cast Away*. At first Hanks struggles to catch any fish using a spear, but over time he succeeds. After enough time spent watching the fish from above, his brain learned their swimming patterns and became better at predicting their next move, eventually allowing him to stop them dead with a spear's throw. This happens in all domains of life; your brain is constantly adapting to the changing world around you and updating its predictions to become more accurate.

Usually this is helpful, but what happens to our predictions in a world with declining social contact? Whether intentional or not, humanity seems to be slipping deeper and deeper into isolation. It has become more common to scroll social media than to put the phone to your ear and call a friend. Instead of chatting with the deli clerk, we have our groceries delivered through an app. We order takeout online when we could grab a table and ask the server about today's specials. Even a nine-to-five workday can be achieved at home, lying in bed with a computer, totally devoid of social contact. These changes have happened slowly, incrementally, and almost without notice over the years.

As a result, I believe that our brains' expectations for social contact have gradually shifted downward. I don't know about you, but I'm no longer shocked when an automated answering service picks up my call. In fact, I'm more surprised when I get a live human. My brain has adjusted to our changing world and updated its predictions accordingly: It now *expects* to talk to a machine rather than a human. This example is a microcosm of what is happening all throughout society. Without really noticing, our brains are adapting to a world where we interact with others less.

To make matters worse, another dramatic shift in 2020 caused our social lives to retreat and shrivel. The COVID-19 pandemic was arguably one of the most isolating events in human existence, plunging billions of people into the murky depths of isolation. While we were stuck at home, the heavy clouds of loneliness rolled in, casting shadows on our well-being and trapping us in an inescapable feeling of distress. Gradually we emerged back into our beloved world, only to find that it had changed drastically. Social interactions were now shrouded in fear and apprehension, carrying the ominous potential that we could be exposed to a strange new illness. Sitting in public spaces carried an unfamiliar layer of anxiety. We didn't want to talk with the butcher anymore, and maybe it was best if they didn't touch our food at all. We adjusted to working, exercising, and even grocery shopping from home. Consequently our brains adapted to seeing our coworkers, gym buddies, and neighbors less. For a species with a highly social brain, *this is bad.*

These many societal changes may have driven our brains to compute new predictions, expecting less and less social contact. But just because the brain has lowered its *expectations* doesn't mean it has lowered its *needs.* As a point of comparison, imagine if similar changes befell our sleeping habits, allowing us to get only three or four hours of sleep a night. What would happen next? Of course we would simply acclimate to this new norm, adjusting our lifestyles and probably drinking way more coffee. We would get used to being more tired, but just because we've adapted *psychologically* to being sleep-deprived doesn't mean that our brains no longer *need* a full eight hours to function optimally. We would be less happy and at higher risk for diseases like Alzheimer's—all because our brains' needs weren't being met. I believe something similar is unfolding in our social lives. A divided world is driving us to withdraw from one another, robbing us of social stimulation that is core to our being. The human brain has a deep and primal need for togetherness, and in a post-interaction world, the brain's ancient social systems are cast into disarray.

To be sure, the space between us truly is growing. From 2013 to 2021, the amount of time Americans spent with friends dropped by about fifteen hours per month, while time spent *alone* rose by over thirty-six hours. According to a 2022 survey, 58 percent of American adults are lonely. In 1990, only 27 percent of Americans reported having three or fewer close friends; by 2021, that number had jumped to 49 percent. Not to mention, the U.S. surgeon general has declared we are stuck in an epidemic of loneliness and isolation. Yes, it's really that bad.

But honestly, does this really matter? Do we really *need* interaction to be happy and healthy? How much does socializing truly impact our well-being?

The answers are yes, yes, and a lot. A wealth of science shows that interactions have a huge influence on our brains. To explain this, we will take a path that surprisingly starts with the *psychology* of interaction before we move on to the underlying neuroscience. Just as my academic journey led me first through psychology and then later to neuroscience, we will take the same route, as I find it helpful for seeing the big picture. Understanding the psychology first establishes a foundation that can then be wrapped in (and explained by) a neuroscientific backing. This will help you comprehend your thoughts and tendencies not just from a behavioral standpoint, but also from *inside the brain* on a cellular level. The interplay between psychology and neuroscience is where the magic really happens. When we can tie actions to molecules, the depth of our knowledge penetrates beyond the superficial, deep into neural worlds unseen and unknown by the naked eye.

## DON'T BE SHY

Ever since we've been around, humans have searched for answers to the question *What makes us happy?* Is it money? Fame? A fast car? Big muscles? A bread bowl full of lobster bisque on a sunny day? Or could it be something as simple as . . . relationships?

It's obvious at this point that I'm going to tell you social interactions are good for you. But you may be surprised by just how potent they are for boosting mood and well-being. Recent research suggests that *being social* and *expressing gratitude* are the most effective ways to boost happiness—even above things like exercising, meditating, and getting out in nature. But what does this really look like in practice? Do we really feel better immediately after interactions?

A study from Washington University in St. Louis suggests so. The researchers asked about 250 college students to wear a wire around campus for one week . . . for science! Of course it wasn't *really* a wire. It was a sound-recording device that the researchers called an "electronically activated recorder" and then cleverly referred to throughout the paper as "the E.A.R." They used it to tape the students' conversations and gather information about their interactions. Okay, I guess it was basically a wire.

During this week, the students also documented how happy they were four times per day. The goal was to see whether having a social interaction influenced their mood. In total, the researchers collected over 150,000 audio recordings, each thirty seconds long. Then an *extremely lucky* group of research assistants got to listen to those recordings and score them. Bless their souls . . . that is nearly fifty-three full days of audio. So what did they find?

Across the board, the students were in a much better mood within an hour after having a conversation, and the more time they had spent talking, the happier they were! The greatest boosts came from interacting with people they liked more and knew better—no surprise there. However, one of the more unexpected findings came from the analysis of the recordings (thank goodness those fifty-three days weren't for nothing). Students showed the strongest bumps in mood when they'd *revealed more about themselves* in conversation. To me, that sounds like a good reason to welcome openness and transparency in our interactions.

It seems clear that *organic* interactions—the ones that happen naturally when we *choose* to socialize—make people happier. But what about interactions that are *not* organic? If the students were instead forced to go interact with a random person, would they enjoy it as much? This question is important because we may need to create new interactions in our daily life if we hope to solve our social problem. Could these nonorganic conversations still boost our mood, or is this only a futile practice?

Luckily Dr. Nicholas Epley, a professor at the University of Chicago, is many steps ahead of us. Epley's lab has studied what happens when people are forced to talk with a stranger, and I'm a huge fan of his research because he runs his studies in *real life*. For example, his team once instructed British commuters riding the train into London to chat with someone random during their trip, to see how it impacted their mood. When all was said and done, the train riders who chatted up a stranger rated their commute more than twice as pleasant as those in the control group,[+] who instead sat in silence. Just like the study carried out with college students, the more time people spent in conversation, the better they rated their commute. The age, gender, or race of the stranger they were talking to didn't matter. The social group always enjoyed their ride more.

It seems that even *forced* interactions can benefit one's mood, and Epley's team has shown this to be true in buses, taxis, and waiting rooms. No matter where people are, they seem to have a better experience when they talk with the strangers around them.

So what does all this research from the field of *psychology* tell us? Here it is in one sentence: Social contact makes people feel good. Whether our interactions happen organically (like bumping into a friend at the store) or are more forced and "artificial" (like approaching a stranger on a train or bus), they always seem to provide a boost in mood. Of course the psychological value of each interaction can differ based on the quality of the interaction and the people we're with (for example, interacting with friends provides the largest boost). However, even *momentary* interactions—like saying thanks to your bus driver as you exit—have been shown to improve well-being. When it comes to lifting your spirits, just a few words muttered in passing may be all it takes.

Over the long term, prioritizing social habits like these can add up to major benefits. Science shows that people who live more social lifestyles are happier, and even *acting* more extroverted can make people feel better. When college students were told to act extroverted for one week (that is, "talkative, assertive, and spontaneous") and introverted for another week ("deliberate, quiet, and reserved"), they showed a big boost in mood while acting extroverted and a strong *decline* in mood while living as an introvert.

With these findings in mind, I encourage you to take a look around your life and see where you can expand socially. Given **Hard Truth No. 1: We live in a divided world**, there are perhaps dozens of missed opportunities per week to seek connection. Instead of letting a family member's call go to voicemail, pick up. When you're tempted to flake on dinner or drinks with friends, consider how that decision could influence your well-being. If we hope to maximize our health, we should commit to prioritizing interaction just as we make pledges and resolutions to eat right, exercise more, and get enough sleep. And seriously—I'm not joking here—try talking to a stranger and see how it feels. I bet you'll be surprised. We are regularly presented with a lovely social fruit that begs to be harvested—a fruit that carries the rare, valuable ability to boost your brain health and mood. So I encourage you to reflect on your life by considering the following questions:

- How often do you see friends and family? How does that compare to years past?

- Do you make isolating decisions like ignoring calls, ordering groceries, or staying in when invited out by a friend?

- Would you feel comfortable starting a conversation with a stranger?

- What decisions could you make to expand your social life and spend more time connecting?

## POP THE HOOD

Before we go any further, it's time to inject some important neuroscience to explain this more deeply. *Why* do social interactions make us feel good in the first place?

First, let's establish a basic principle: For every *psychological* experience you have, there is a corresponding event *in your brain* driving that experience. This is precisely the magical interplay between psychology and neuroscience that I alluded to earlier. For instance, when you experience a sense of joy while petting a dog, certain parts of your brain are revving up to create that feeling of joy (probably areas that process rewards). If the dog were to turn around and fart noisily in your direction, those reward areas might turn down their activity, resulting in a bit *less* joy. Meanwhile, other regions may turn on—like those involved in disgust. Understanding the gears turning inside you may deepen your appreciation for the sensations you experience. If you think about the human body like a car, the brain is the engine that drives us. Sometimes it's valuable to pop the hood and take a look.

Now let's apply this principle to social interactions. In the first part of this chapter, we established our psychological phenomenon: that **social interactions make people feel good**. Now we must ask: **What's happening in the brain?** To answer this, we will turn to the neuroscience of *social reward*—a concept with deep roots in the ancient history of humankind . . .

⊕

Through my years of studying science, I've come to appreciate that almost every feature of the body has some cool evolutionary origin. For example, isn't it odd that you have a nose? I know, weird question to ask, but really, does that *need* to be there? It's not exactly the most attractive thing. Of course you need nostrils so you can breathe when your mouth is occupied (like when you're eating), but why the whole nose? Couldn't you just have two nostril *holes* on your face like Voldemort? The nose actually has many purposes. For one, it functions as a filter, a trap lined with hair and mucus to prevent unwanted things from entering your airway. Second, it ensures that your nostrils are pointed downward, to prevent water and debris from constantly dripping into them. As clunky as noses are, we have them because they help us survive. We can only hope that Voldemort has a strong immune system to protect him from all the shit that probably gets into his face.

Similarly, the brain's inner workings have purposeful origins rooted in evolution. Just like there's a good explanation for why we have a nose, there's also a good reason why interaction makes the brain feel good. It boils down to one simple fact: In an ancient world, *we survived better in groups*. Thousands of centuries ago, being alone would've totally sucked. Imagine taking on a saber-toothed tiger in a one-on-one battle. It sounds like certain death (and not a pleasant one). But with fifteen or twenty compatriots, you might have a chance. When it came down to "survival of the fittest," the *fittest* humans were often the most *social* humans. We may not be the strongest or most vicious beasts on Earth, but we're damn smart, and we can communicate incredibly well. That makes us dangerously effective when we work in teams. So for the sake of our survival, we were wired to be social.

But what does that really mean, to be "wired to be social"? Well, a little exercise might help. Let's pretend you were born to work for the Earth and serve the many species that live here—a majestic fate. Your job is to make sure all animals stay alive, and you do this by selecting the helpful adaptations that will promote their survival. Your boss, Mother Nature herself, has given you a new assignment: *humans*. You must keep them alive by shaping them however is necessary. As you observe them, you notice that they're exceptional at hunting and defending themselves when they work together. So to keep them alive, you decide to design a new feature that makes them *want* to exist in groups. You can tinker with their brains or bodies however you want. What feature would you select?

There is a correct answer to this. At least, there was a specific adaptation that came about in reality (and it wasn't that unsightly nose). It was a simple but clever solution: *social reward*. Our brains' chemical systems were etched so that interacting with others would *feel pleasant*. Thanks to this, humans were naturally inclined to stick together. While this adaptation may have happened many millennia ago, these social reward systems remain in our brains today and continue to play a key role in shaping today's societies.

I mean, just think about it. In modern life, how do people typically spend their weekends? Aside from catching up on chores and "rotting" on the couch, many people strive to dedicate this cherished free time to being with others. We visit with family or friends and gather in public areas. On our precious Friday and Saturday nights, we pile into cramped rooms together and drink alcohol to ease our interactions. We could be doing *anything* with these valuable hours, but we often want to spend them in this strange social stew. Interesting, isn't it? To me, it's a clear and beautiful expression of our innate drive for togetherness—a by-product of our evolutionary need to latch together for survival.

However, these social instincts may be challenged in today's world thanks to the many new forms of social reward that contest our innate drive to be together. For example, you can upload a photo on social media at any time and begin collecting likes from other users. This is a clear gamification of social reward—an attempt to take advantage of your brain's natural desire to gain approval from your community. Your brain's social systems are tickled by others pouring positive attention into you with the tap of a button, making these apps hard to put away. When you turn on the television, you can watch shows or movies where characters fall in love, obtain high social status, or defeat a shared enemy—all of which captivate the brain's social reward systems. Maybe the reason why it's *so easy* to ditch a night out in favor of watching TV or scrolling social media is because we can still obtain a bit of that social reward through our devices. Perhaps we've become content with substituting these synthetic social rewards for the real deal, and consequently grown less inclined to connect face-to-face.

We shouldn't take our social nature for granted. After all, not all animals are born social like we are. For instance, *solitary* animals like tigers prefer living alone. Why? Well, tigers are perfectly capable of killing prey on their own, so teaming up would just mean more mouths to split dinner with. With there being no evolutionary pressure to exist in groups, their brains didn't come to reward them for being together. On the other hand, there are loads of other species that *do* share our drive to be social, because they survive best in groups. *Monkeys* travel in troops, *fish* swim in schools, *dogs* hunt in packs, and even *mice* live in colonies. Because of this likeness to our brains, much of the neuroscience research on social behavior is done with mice. One clear example of social reward in mice comes from some of my own research.

In the first year I spent working on my PhD, I had a lot of cool ideas (probably because my brain worked a lot better back then). I found myself thinking about how people sometimes do idiotic things for social reasons. For instance, are you guilty of spending too much money on dinner to impress a date? Have you gotten sick after you and your friends attended a concert in cold, wet weather? What about staying up late into the evening to stick around the bar? I've done all three, and probably worse. Humans are often willing to put ourselves through uncomfortable things to be with others, and that's presumably because of the social reward system in our brain. Since mice are socially motivated like us, I wondered if they would do this type of stuff, too.

One day in the lab, I spun my desk chair around to my colleague Dr. Zijun Wang and started talking about this. How might a researcher test a mouse's willingness to do idiotic things for social reasons? Mice don't go on dates, have bedtimes, or hold concerts (but if they did, I would definitely want to see one). Thanks to my young, fresh brain (and Dr. Wang's help), an idea emerged.

Imagine a platform shaped like a plus sign that stands a few feet off the ground. Two of the four arms are sheltered by walls, while the other two are completely exposed. This device is called the *elevated plus maze*, and scientists use it to test anxiety in mice (yes, we can do that). Those wide-open unprotected arms are terrifying if you're a mouse, while the closed arms are a shadowy, protected sanctuary—a mouse's dream. Because of this, mice spend almost all their time in the closed arms, venturing into the open arms only in a few courageous bursts. The less time they spend exploring those scary open arms, the more anxious we assume they are. Dr. Wang and I saw this as the perfect venue to measure social motivation. What if we put a second mouse in one of the open arms? Would your average mouse be willing to go out on a limb (literally) just to pay a buddy a visit? Aha! We had found our mouse equivalent of a cold, rainy concert. Now, time to see if the mice were willing to get uncomfy to be social . . .

Amazingly, it worked. When we put a mouse at the end of one open arm, the free-running mouse spent about four times longer

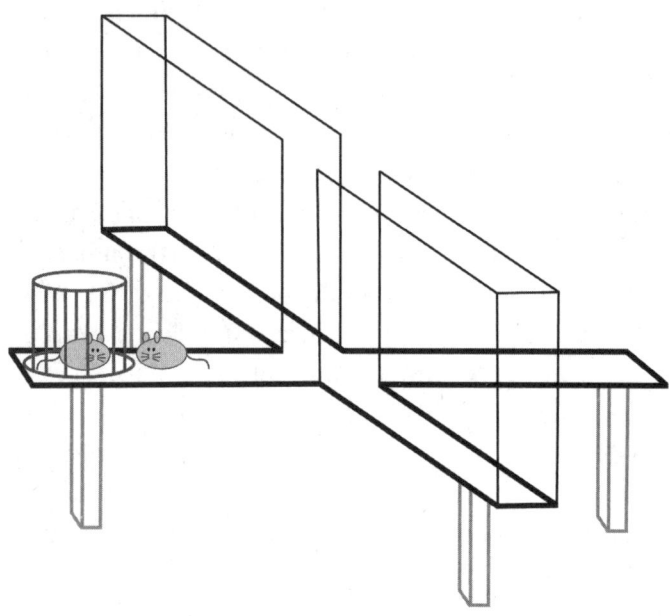

out there. They were *four times more willing* to brave the frightening conditions if it meant they could have a social interaction. In contrast, we found that mice with autism-related gene mutations showed no such social motivation—they were hardly willing to enter the open arm for a social interaction. (By the way, I did publish the figure above in a scientific journal, complete with the cute little oval mouse. I privately named him Tony.)

Clearly, interactions are rewarding for mice. In fact, they will press a lever up to twenty times just to hang out with another mouse for fifteen seconds. Because of this similarity to humans, we can study their brains to figure out *how* social reward works— and that is exactly what neuroscientists have done. This is where we get the answer to our big question: What's happening in the brain to create social reward?

It turns out that there is not a single brain chemical that controls social reward. Rather, it involves three key players: *oxytocin, serotonin,* and *dopamine.* You might recognize these three chemicals, as they are some of the more popular *neurotransmitters,* or chemical signals that brain cells pass to one another to communicate. (*Psst:* If you're looking for an in-depth refresher on how brain signaling⁺ works, check out the corresponding entry in the appendix.)

But to be honest, who cares? Does knowing the name of three chemicals that make interactions feel good *really* help you understand yourself? I don't think so. At least not until we explain some more about what these chemical signals do and where in the brain they're acting.

Let's start with oxytocin, which is often called the "love hormone" because of its role in social bonding and attachment. Most of the brain's oxytocin is produced in the *paraventricular nucleus.* This is a cluster of neurons named for its position next to one of the brain's fluid-filled cavities, which are called *ventricles.* You can think of this brain area as an oxytocin factory sitting beside a beautiful pond of cerebrospinal fluid. This factory produces and packages oxytocin, then ships it off to other brain areas.

Research shows that when this oxytocin is sent to a specific brain area called the *nucleus accumbens,* it creates that sense of social reward. The nucleus accumbens is an important brain area to remember, because it will come up many times throughout this book (so pay attention!). It's made up of an outer shell wrapped around an internal core, sort of like an M&M's candy. Coincidentally, it's also almost exactly the size of an M&M, but with a more stretched-out shape and a flattened top. It's found on both sides of the brain, embedded a few inches behind the eyes (that makes two M&Ms, one on each side). These M&Ms are very well connected, receiving wirelike projections from many other regions, sort of like a central train station in a large city. This says a lot about how important the nucleus accumbens is; like a CEO whose phone never stops ringing as she handles a stream of business deals, the nucleus accumbens is one of the brain's chief operators. Specifically, it gathers information from multiple brain systems to shape our *motivations,* with a major goal of pursuing rewards. For instance, when you're trying to decide something like "Should I take another bite of this pizza?" your nucleus accumbens might chime in, saying, "Yes, it tastes delicious . . . and just *look* at that crust!" Or when you enter a casino, your nucleus accumbens might say, "Go to the blackjack tables—remember how much you won last time?" In both scenarios, it's shaping your next move by driving you to collect a reward. Considering that this is what the nucleus accumbens does, it's quite fitting to think of it like a sweet, chocolatey M&M. When you feel the urge to reach for another crunchy, sugary handful, it's likely to be your nucleus accumbens motivating that behavior.

Seeing as the nucleus accumbens is so tied up in processing rewards, it makes perfect sense that it would handle *social* rewards, too. Research shows that the brain achieves this by sending oxytocin there, but the oxytocin doesn't act alone. When that oxytocin pours in, it causes *more serotonin to be released,* and this serotonin is also critical. When you're having a good conversation with a friend or cuddling with your spouse, it's very likely that oxytocin and serotonin are flowing in your nucleus accumbens. By working together in this reward-driven brain region, the two neurotransmitters help brains appreciate the joy of social connection.

Now, we can't forget about the third member of this party: dopamine. Surely you've heard of it, the neurotransmitter that's famous for its role in reward and motivation. However, dopamine is often misunderstood. People tend to think of dopamine as a "happy hormone," but it's so much more than just a pleasure chemical. Dopamine is a molecule of *learning*. It's released when you experience things that are pleasant or good for you—like taking a hot shower, eating a delicious meal, or tucking yourself into a buttery-soft blanket—to help your brain remember that these experiences are pleasant. As a result, you will be more motivated to take another hot shower, eat that same meal again, or use that same blanket the next time you have the chance. Dopamine is thus responsible for *tracking* the value of rewards and *motivating* your behavior accordingly. In other words, it's the brain's built-in reinforcement system. It's why rats will press a lever relentlessly to get cocaine (and humans will do much more): because cocaine directly stimulates dopamine signaling. This makes the drug intensely reinforcing, as it sends a powerful signal in your brain to "go back for more." Dopamine is so reinforcing that rats will literally press a lever just to have their brain's dopamine centers stimulated directly. No need for drugs when you can just get zapped right in your dopamine factory!

So how does dopamine tie into social reward? Research suggests that oxytocin can *stimulate* the release of dopamine. This presumably makes social interactions reinforcing, to keep us coming back. It's probably the reason why humans attend rainy concerts to be with friends, and why mice will enter a scary environment to sniff another mouse: because this dopamine makes our brains inherently *motivated* to seek social contact.

The bottom line is that **social interactions make the brain release rewarding chemicals**. Oxytocin—which has some rewarding properties itself—starts flowing when we connect with others. This drives the release of serotonin and dopamine, two other neurotransmitters linked to pleasure and reward. This is sort of like a set of three dominoes, where the fall of one big domino (oxytocin) triggers the fall of two additional dominoes (serotonin and dopamine). The release of these three neurotransmitters at once creates an enchanting concoction of neurochemicals that makes us feel outstanding. In fact, the combination of serotonin *and* dopamine might be especially powerful, as one of the only drugs to have the same effect in the brain is MDMA. Also known as *molly* or *ecstasy*, the drug is known for creating a powerful sense of euphoria and social connection. The fact that oxytocin influences the very same brain systems says a lot.

It seems we have our answer to the question we've been searching for, and an explanation for *why interactions make us feel good*: The chemical signature of social interactions in the brain is one characterized by pleasure and reward! If we think back to our evolutionary history, this makes perfect sense. Our brains are built so that social connection feels good and is inherently reinforcing. This keeps us together—and therefore keeps us alive. *This* is what it means to be "wired for connection."

Earlier on, I mentioned that mice are commonly used in social neuroscience because just like us they're motivated to exist in groups. The truth is, almost everything I just described was first discovered in the brains of rodents, not humans. However, those three neurotransmitters have also been linked to social reward in our brains (I'm assuming you're a human, not a mouse). When people squirt oxytocin up into their nose—a surprisingly common way to deliver oxytocin—they show increased activity in reward-related brain areas and find human touch more pleasant. As for serotonin, people perform worse at a social reward–learning task after their brain's serotonin has been depleted. Dopamine has been linked to social reward through extroversion: People with certain variants of dopamine receptor genes are more prone to extroversion. A relationship between dopamine and extroversion would make perfect sense. If certain people's brains release more dopamine when they socialize, those people would be more motivated to pursue social interaction again, just like dopamine motivates you to get back in the hot shower or tuck into that buttery-soft blanket. As a result, those people would be more extroverted. It turns out I was onto something with my childhood prediction of a social continuum: Each person's unique social preferences really *do* reflect differences in how their brain functions.

The fact that our brains have a built-in system for social reward says it all: We are meant to be together. This system is the reason why we feel good after talking with a stranger on the train. It's why we often choose to spend our weekends visiting with friends and family. It's why we should pick up the phone when it buzzes on our desk. Because when we engage in social activities, we bring online neural systems that drive a sense of reward and gratification.

This brings me back to **Hard Truth No. 1: We live in a divided world.** Although we may be veering off track as a species (that is, pulling away from one another), there's good news: The brain will always appreciate company. Each of us alive today is standing on the shoulders of *hundreds of thousands of years* of social precedent... perhaps even millions. Long before history was documented, our ancestors lived social lives. This means we aren't just wired for togetherness; we are *hard*wired for it. Just as your brain demands a full night of sleep, it also needs a sufficient helping of social interactions. Even when our culture fails to prioritize connection, our brains always will. Because at one point long ago—well before the advent of things like social media and remote work—it simply had no choice. Without social connection, we may have died off.

On the other hand, what happens when we allow division to win? In a modern world that's creating more and more space between us, what happens if we fail to come together? To gather a complete picture of social interaction, we must flip this problem over and stare at the dirty underside of isolation. In the end, you may be surprised by what's at stake when we deprive our brains of connection.

## Key Takeaways

1. As societal changes have driven us toward seclusion, our predictive brains have come to expect less interaction. However, shifted predictions do not equate to shifted needs.

2. Being social is one of the most effective ways to boost happiness. Even a short conversation with a stranger can spike your mood and elevate your experiences.

3. In ancient times, humans needed to stick together to survive. As a result, our brains rewarded us for existing in groups. This *social reward system* remains

in our brains today and allows us to derive joy from socializing.

4. The brain's social reward system is driven by oxytocin, dopamine, and serotonin. These neurochemicals make socializing reinforcing, which makes us feel good and want to go back for more.

To view the references cited in this chapter, please visit benrein .com/book.

# 2

# THE ONLY ORGAN THAT GETS LONELY

## How Social Isolation Dismantles Brain Function

Imagine a reality where you live in complete isolation. Suppose that everything in your current life remains the same—your home, your job, your bed—but you have *no* social contact. You are the last human remaining on an Earth that keeps on spinning.

Each morning you awaken to a silent, deserted home. You take barren roads to work, with no traffic to contend with. Your car ride is entirely quiet, as there are no voices on the radio to listen to, nor is there any music to sing along with. Trains without passengers roll past you. You arrive at an empty workplace and sit at your desk, prepared for a day of working alone. Your lunch breaks are spent strolling the abandoned halls. In the evenings, you cook dinner for one. Weekends are just the same, filled with unaccompanied chores or lonely walks through the park.

Initially your loneliness might feel like boredom. Perhaps you would notice how dull and repetitive your thoughts are without input from others. Over time, you might gradually settle into a shadowy pit of solitude—not pleasant, but tolerable for now. But over the following months, those feelings would deepen until they became unignorable. The brittle floor of the pit would erode, exposing a deep crater of discomfort and delirium as you long for social connection. Under the pressure of isolation, your physical and mental health would progressively crumble.

How does this reality feel? My guess is it feels bad. I personally find it *stressful*; my chest tightens at the thought of such a lonely existence. However, this imaginary world isn't so imaginary for some. Prisoners held in solitary confinement live it every day. For widowed seniors or ostracized students, this vision isn't far from reality. That's not just a terrible tragedy, but a serious public health concern.

Humans realized that isolation is agonizing a long time ago. All the way back in the 1700s, the controversial practice of solitary confinement was already being used to punish prisoners. It is one of the worst fates a human can experience and is deliberately intended to cause suffering. Some of the earliest American attempts produced horrible, gruesome results. In the early 1820s, New York State passed legislation allowing Auburn State Prison to experiment with solitary confinement. The prison's warden recounts how the experiment affected prisoners: "one was so desperate that he sprang from his cell . . . and threw himself from the gallery upon the pavement . . . another beat and mangled his head against the walls of his cell until he destroyed one of his eyes." These prisoners' behaviors paint a picture of pure desperation. It's chilling that they seemed to regard death as a favorable alternative to living without social contact.

More recent reports show that long periods of solitary confinement can trigger vivid multisensory hallucinations, hypersensitivity to noises, and aimless violence. A study of over two hundred thousand prisoners found that those who were isolated for *any* amount of time during their sentence were 24 percent more likely to die in the year after their release. They were 78 percent more likely to die by suicide.

Why does solitary confinement have such terrible consequences? Perhaps it's because we were evolutionarily wired to be together. Social connection, like sleep or nutrition, is a basic need of the human brain. Being held in isolation goes against our deepest and most primal instincts to be around others. It buries a seed of primordial distress within us, which may swell into madness if it's allowed to grow and take root. We are exquisite creatures—packed head to toe with impressive machinery—but unlike the many other organs within us, the brain gets lonely. This is a unique property. Day and night, your *heart* thumps out nutrients to your hungry tissues, your *intestines* churn food along their long winding path, and your *diaphragm* works tirelessly to fill your lungs with air. These organs simply *work*, exerting themselves endlessly, without question. However, the brain is different. It has needs that go beyond simple nutritional demands like oxygen and glucose. Unlike the rest of our organs, the squishy wrinkled mass that holds our every thought, feeling, and secret needs *companionship*. It craves the company of others. Without this, it fails.

Or, at least, it struggles. Social bonds are a vital ingredient for brain health, and what we may be missing is that brain health underlies *essentially every bit of our experience*. "Having a healthy brain" doesn't just mean solving puzzles quickly or staying focused on tasks; it's also about how well you manage your emotions, how sleepy you feel at work, how much exercise you can tolerate, how hungry you are, and so much more. It's common to believe that the key to bolstering brain function is eating a special diet, taking a certain supplement, or doing crossword puzzles, but these elements are only part of the picture. To nurture our brains, we should begin by ensuring that we're meeting their most basic needs, and social contact is among those.

Solitary confinement is probably the most extreme example of loneliness, but it reveals what many of us already know to be true: Being alone hurts. For the average person, isolation comes in spotty waves or short bouts: a day or two without a good conversation, or a few days in the house for those who live alone. It looks like a remote worker in a studio apartment letting a workweek slip by without seeing anyone in person. It looks like a student stuck at home with the flu on the day of prom. It looks like a patient recovering from surgery on the couch while their partner goes into the office. These relatively brief periods aren't nearly enough to drive someone to madness (though it can feel that way sometimes), but that doesn't mean they're negligible. Even at shorter lengths, solitude can seriously impact well-being.

Studies on the recovery of hospital patients offer a clear example of how short-term isolation can wear us down. When patients are secluded during their stay (often done to prevent the spread of contagious diseases), they suffer worse depression and anxiety, and they are at a higher risk of experiencing adverse events. If the isolation continues afterward, things can get even worse. When patients are discharged from the hospital after a heart attack, those who live alone are more than *twice* as likely to die in the next three years. In contrast, those who receive the strongest social support after a stroke show the greatest recovery.

Another glimpse comes from the pandemic lockdowns of 2020, which offered a heartbreaking but fascinating look at the result of loneliness. Studies have shown that social restrictions were associated with higher levels of depression, stress, and loneliness, and Americans who sheltered alone struggled more than those who were stuck with family members. Meanwhile, the longer people were isolated, the worse their depression was.

Despite the consequences of isolation, it seems we are failing to prioritize connection in modern life. We continue to live in a divided world, and while we fail to make change, the risks remain very real: Studies tracking millions of people show that isolation is associated with a 32 percent higher risk of death *by any cause*. How do we move forward and extract ourselves from this dangerous trap?

I wonder if perhaps we're not armed with the right information— attempting to navigate this confusing and scary reality in a knowledge deficit. If we fully understood what's truly at stake, maybe we would work harder to shake our way to safety. What does isolation do to the brain? Why does being alone literally *increase the risk of death*? It's time to find out *why brains need friends*.

## THE DEADLIEST NUMBER

In 2015, Alzheimer's and stroke researcher Dr. Joshua Crapser published a scientific paper about how isolation impacts brain health. The paper's provocative title said it all: **"One Is the Deadliest Number."**

While conducting research on brain *aging,* Dr. Crapser and his colleagues discovered something shocking. They were studying strokes, which happen when blood stops flowing to certain parts of the brain, causing neurons to be starved of oxygen and die off. The lab would artificially induce strokes in mice by temporarily blocking an artery that supplies blood to the brain. Their original goal was to figure out how the brain recovers from such a devastating event, but they got sidetracked when they noticed a surprising trend in their results. For some reason, mice that lived *alone* had much more severe strokes! Somehow, the exact same stroke was leaving behind *more* damage in their brains, and a larger area of starved, dead cells. What's more, the isolated mice showed worse symptoms, took longer to recover, and were more likely to die.

This seemed farfetched. How could we be sure that their social conditions are truly to blame? Well, other studies have found that when mice are only *partially* isolated—separated by a clear barrier where they can see each other but can't make physical contact—their strokes fall somewhere in the middle. They show more damage than mice living in groups, but less than those fully isolated.

I find this completely astounding. I mean seriously, how is this even possible? This implies that *socializing* somehow protects neurons from *suffocation*. If two people suffered the exact same stroke (suppose their brains were starved of blood for exactly five minutes), whoever has a less active social life would theoretically have more brain damage. How can this be?

Recall the exercise at the beginning of this chapter. Did the thought of a life without social contact make you feel *stressed*? Does your chest feel tight at the idea of never seeing your loved ones again or of being held in solitary confinement for long periods? If so, that's entirely normal. Because social isolation *is* in fact a form of stress, and the body treats it as such. Considering our evolutionary history, this makes perfect sense. Perhaps in our ancient world this distress signal increased our survival chances by motivating us to seek community. Now, millennia later, that tightness you feel in your chest may be an artifact of this prehistoric survival system. However, that doesn't mean it's still good for us today. Through specific mechanisms in the brain, this stress could be leading to the exaggerated strokes seen in isolated mice.

The human body responds to stress in specific, predictable ways. For instance, the *HPA axis*—a complicated system that involves many parts and pieces—is one of the brain's primary stress-response systems. When you're stressed, a neurotransmitter and hormone called *norepinephrine* (aka noradrenaline) is released to drive that fight-or-flight response we all know so well. This wakes up a brain region called the *hypothalamus*, which is generally responsible for balancing the body and maintaining homeostasis; think of things like regulating body temperature and heart rate. When the hypothalamus notices these alarm bells ringing, it releases a hormone called *corticotropin-releasing factor*, which notifies the *pituitary gland* to discharge another hormone into the bloodstream called *adrenocorticotropic hormone*. Finally, this causes the *adrenal glands* to release *glucocorticoids*. If you were wondering what the HPA axis stood for, it's *hypothalamic-pituitary-adrenocortical axis*. Now you can see where that name comes from—and how ridiculously complex the stress response is.

Don't worry—there's no need to memorize all that. The part I want to focus on is the last bit, about glucocorticoids. These are hormones that act on many tissues in the body to initiate a stress response. In humans, the primary glucocorticoid goes by a familiar name: *cortisol*—often referred to as the "stress hormone." Cortisol prepares the body to take on whatever threat is causing the stress. The body thinks a fight is coming, and it's readying the troops. That means increasing the heart rate, suppressing nonessential bodily functions like digestion, and increasing blood sugar to feed the muscles and tissues. Importantly, one of the other things cortisol does is reduce inflammation. That's because when you're facing something stressful like being chased by a predator, inflammation does you no good. Incidentally, if you're wondering why the word *glucocorticoid* sounds familiar, you may have been prescribed one (like prednisone) for the same reason: to reduce inflammation in your body.

Wait a minute . . . if cortisol is anti-inflammatory, wouldn't stress be a good thing?

On paper, it may sound that way. But there's a catch. When stress shifts from being short term (often called *acute* stress) to long term (*chronic* stress), bad things happen. Chronic stress causes cortisol levels to stay elevated for a long time, which can lead your body to become desensitized to it. Just like we can't stay in fight-or-flight mode forever, we can't expect our body's tissues to remain in this stress-response state indefinitely. Eventually they stop responding to the cortisol, no longer letting it reduce their inflammation. As a result, when the body is under chronic stress, this hormone can lose its anti-inflammatory properties. This is a problem. With one of the main anti-inflammatory systems out of order, the body becomes vulnerable to chronic inflammation. And that, my friend, is a bane to health and wellness. Chronic inflammation can cause damage to healthy tissues in the body, and is associated with conditions like heart disease, diabetes, and cancer.

This seems to be what's happening in social isolation. Being that it's a form of stress, isolated people show higher cortisol levels, and those with fewer close relationships have higher cortisol output. One study conducted at NASA's Johnson Space Center in Houston had subjects isolate for thirty days inside a 650-square-foot enclosure. Just one week into the experiment, their cortisol levels had increased by 56 percent on average. This may help explain why isolated hospital patients suffer worse anxiety and depression, and why a week of uninterrupted remote work can be unsettling. The subjects' cortisol levels remained high for the full thirty days, only recovering a few days after they returned to normal life. Let me say that again. Cortisol levels rose during isolation and dropped after a return to socialization. Considering that just about everyone wants to protect their health and well-being, are we neglecting a hero in plain sight? Social contact is free, pleasant, and effective. Meanwhile, there are thousands of fad diets, supplements, and other expensive elixirs out there claiming to heal our cortisol levels. Perhaps we can simply turn to a more straightforward (and cheaper) solution: one another.

Is chronic inflammation the true culprit here? Is it the reason why isolated people are more likely to die by any cause, and why Dr. Crapser's isolated mice had worse strokes? Research he led with University of Texas Health Science Center at Houston professor Dr. Louise McCullough suggests so. When they suppressed inflammation in their isolated mice, it protected them—their strokes were no worse than those of the socialized mice. It seems that when isolation causes inflammation, the brain becomes less resilient to injuries like strokes.

Another example of this comes not from the brain, but from the heart. Loneliness is associated with an increased risk of heart disease, and people who live alone are more likely to die after a heart attack. However, when mice are given oxytocin—that key neurotransmitter released during interactions—it appears to protect them from the negative effects of isolation on heart health. This is presumably because oxytocin doesn't just drive social reward and bonding but also has *anti-inflammatory* properties. The significance of this cannot be overstated. It suggests that interactions don't just make us feel good, they literally protect us against disease and death in many forms by battling dangerous inflammation within us. Recognizing this, we can start to appreciate the true grip that social connection has on our health. We are beginning to see the reality of **Hard Truth No. 2: Division is the enemy of brain health** (not to mention the health of the rest of our bodies, too).

Being deprived of anything you need is distressing, and social nourishment is no different. Humans are social animals; interaction is inherent to our brain function. If we weren't socially inclined, isolation probably wouldn't have these negative effects. In fact, those solitary animals that prefer living alone (like tigers or lemurs) *do not* show increased cortisol when they are isolated. Instead, they show increased cortisol when forced to live with others! Their social needs are the opposite of ours, so their stress systems are wired accordingly.

In early 2024, my mom suffered a devastating bout of depression as she coped with gradually losing her mother to Alzheimer's disease. Her mood was dark for weeks, and as she slumped further into despair, she found herself losing the desire to socialize. It just didn't seem worthwhile to leave the house or see friends; however, this self-isolation was quietly causing her depression to worsen. When a lifelong friend invited her to dinner, my mom declined. Luckily, her friend could see that my mom was depressed and needed friendship more than ever. "I don't care, you have to come!" she told her. Not wanting to disappoint her friend, my mom decided to go. She couldn't believe the way her body responded. That night she slept terrifically, and she woke up the next day with an all-new attitude. After just a few hours with friends, she found her mood uplifted. Had her friend not insisted on spending time together, who knows how this sustained isolation could have impacted my mom's health.

Not only do we experience great pleasure and reward when we're with others, but we experience distress and health consequences when we're apart. It's a powerful biological imperative, and one we should pay attention to. Even when we don't feel up to socializing, our brains and bodies stand to benefit.

As a word of caution, it would be wrong of me to merely equate isolation to "stress" in the brain and say no more. The truth is that isolation is complex. There are many other things happening inside a lonely brain besides inflammation. When people are lonely, their brain starts to process social information in a new, unhelpful way. They seem to focus more on the *bad* in their interactions. They show bigger brain responses in the visual cortex when looking at negative social images such as a man slapping a woman. Lonelier people also struggle with trust, showing weaker activity in trust-related brain areas like the anterior insula, nucleus accumbens, and amygdala. Unluckily, lonelier people are also rated as less trustworthy by *others*.

In our divided modern world, I can't help but wonder if the loneliness we've allowed to manifest is driving us even *further* apart. Maybe our isolation is making us distrust one another more, enhancing our vulnerability to things like political polarization, culture wars, and other factors that widen our divides. For a society that was already spiraling into distrust in the late 2010s over rising political intensity and other cultural factors, the COVID-19 pandemic may have been the worst thing that could have happened.

To make matters worse, being isolated can make socializing less enjoyable, which may make it difficult to escape a state of loneliness. Lonely people experience *less social reward* and show reduced activity in part of the brain's reward circuitry—the ventral striatum—when they view positive social images. As a result, lonely people experience weaker boosts in mood after pleasant interactions and show lower oxytocin levels. This all carries an important lesson: When you're feeling lonely and isolated, it may take a few interactions to get you out of your shell and settled back into your routine. My mom was lucky that a single night out impacted her so much, but the effects may not always be so immediate. If the first time visiting with friends or family doesn't do the trick or you don't feel like yourself, give it time. Your loneliness may literally be altering your brain function in ways that make it difficult for you to find meaningful connection.

There's also something else we must address: The effects of isolation can differ depending on one's age, and it matters *when* isolation strikes. In this regard, there are two populations we should be paying special attention to: *children* and the *elderly*. These groups seem to be vulnerable to isolation in unique ways that are important to understand. Let's start where it all begins: childhood.

## THE IMPORTANCE OF SOCIAL EXPERIENCE IN EARLY LIFE

In 1957, a little girl was born in California named Susan Wiley . . . but thirteen years later, the world would learn of her by a different name: Genie.

Genie was cursed with a childhood that nobody should ever experience. For most of her early life, she was the victim of severe parental neglect and abuse. Almost all her time was spent locked in a room alone, undernourished, and immobilized in a makeshift harness. Her life was devoid of social experience. She was hardly exposed to any speech, as her only interactions came in a twisted form from her abusive father. Reports state that if Genie made any noises, her father would bark and growl at her like a dog, beating and scratching her to keep her quiet. In late 1970, she was thankfully discovered by California authorities when she entered a welfare office with her mother. The social workers could tell something was wrong just from looking at her, and initiated an investigation. Soon her living conditions came to light, and the news stories began to break.

Genie was described as "feral." Her behaviors were symptomatic of a life without typical healthy human contact. She had virtually no language skills and recognized only about twenty words. She could express herself only nonverbally, and her communication was not very effective, involving no facial expressions or body language. Genie was indifferent to the presence of others, often walking away from people as they spoke to her. After being taken into care by the state of California, she continued to struggle with language and never fully acquired the ability to communicate through speech. Much of the damage from her early life was unfortunately permanent.

As tragic as Genie's story is, it has become a landmark case in psychology, offering a rare glimpse into what human biology could resemble without any exposure to social stimuli. Genie's case peeks into a dark, forbidden corner and shines light on the question: What does a child develop into without *any* social interaction? What we find illuminated there is jarring and disturbing.

Without social contact, Genie simply didn't acquire skills for interaction. The mannerisms we tend to engage in—like looking at the person talking, expressing emotions through facial expressions, or laughing at the appropriate times in a story—are seemingly *learned*. Having never practiced social exchange, Genie lacked these fundamental elements. This suggests that these behaviors aren't ingrained in our genes; they must be trained and learned through experience.

Thankfully, the brain is built for this due to its high levels of *plasticity*, which refers to its ability to change and adapt over time. By that, I don't mean that it can grow new areas or change shape as you learn things; rather, this plasticity refers to tiny changes happening at the *synapses*. A synapse is a microscopic junction between two brain cells where they can interact by sending and receiving neurotransmitters. (If this is new information for you, I'd encourage you to go check out the "brain signaling" section in the appendix!) Each of the approximately 86 billion neurons in your brain has thousands of synapses, amounting to *trillions* of synaptic junctions. As neuroscientist Dr. David Eagleman once put it, "If you took just a cubic centimeter of the brain, there are more connections in there than there are stars in the Milky Way galaxy."

Even more impressive, every one of these synapses can change. When two brain cells interact frequently, the synapses between them may get stronger. If left inactive, synapses may weaken or fall apart entirely. In many ways, each synapse is like a little muscle that grows with repeated use or will atrophy if neglected. When you do bicep curls every day, your arms will grow bigger because they stack on muscle to make this motion easier next time. Similarly, if you practice reciting poetry daily, you will exercise the brain systems that control these language skills, causing the synapses there to grow stronger. As a consequence, you will become better at reciting poetry—just like your reinforced biceps have less trouble lifting that barbell. This plasticity is what allows us to learn, grow, and adapt to changing environments.

All this change is not cheap. It takes a lot of energy for the brain to keep up with all this. Just think how costly it would be to have construction crews constantly deployed on every road in a big city. In early life the brain is incredibly plastic, as it's important for children to be able to soak up information as effortlessly as a sponge absorbs water. However, this capacity for learning and adapting declines with age—largely because of how expensive it is. By the time we are seventy or eighty, our brain is much less plastic. By this point, the brain has seven or eight decades of experience, making it a well-trained *prediction machine*. It doesn't need to be as adaptive to learning new things, and besides, the body is better off using that energy elsewhere, like powering our aged muscles. That's not to say *all* plasticity is gone by the time we meet our grandchildren. Of course, the brain can still learn and adapt; it's just not as spry as it used to be. It would be wrong to think that making changes to our habits in late life is futile. The brain is still receptive to change, and even senior citizens can respond robustly to new habits.

The extreme brain plasticity we experience in early life is meant to help us figure out the world around us. Through the first few years of development, the brain goes through specific phases of intense plasticity called *sensitive periods*, in which certain skills can be acquired very easily. During sensitive periods, specific brain circuits become highly flexible for a short time, allowing them to be heavily shaped by experience. For example, have you ever noticed how easily kids can pick up a language? That's because there's a sensitive period for language acquisition in childhood: The brain areas guiding language become super plastic, expanding the brain's ability to take in new information.

But that's not all. The brain also seems to have a *social* sensitive period in the first five to ten years of life, when it is especially sensitive to learning social skills through interaction. If a child is isolated during this period as Genie was, their social skills may suffer. Unfortunately, Genie may have simply missed this window of plasticity due to her isolation. "She was smart," recalled Dr. Susan Curtiss, a former professor of linguistics at UCLA who studied Genie extensively. "She had other signs of intelligence." But when it came to socializing, Genie struggled. By the time she was rescued, it was too late: The period of heightened social plasticity had come and gone.

Genie's story is extremely rare and unique. It reveals what happens to a developing brain with *no* social contact at all. However, there is plenty of other evidence showing how important social experience is in early life. Children raised in orphanages with minimal social contact show developmental impairment. Infants who were born during COVID-19 and spent their first year in quarantines show poorer communication skills than those born before the pandemic, along with reduced verbal, cognitive, and motor skills. Animal studies suggest the same. Monkeys isolated for six or more months in early life suffer irreversible damage to their social interactions, often failing to develop proper play behaviors. When rats are isolated as pups, they may struggle with recognizing other rats as adults. Even dogs show evidence of a social sensitive period: When puppies are isolated from humans during the first fourteen weeks of life, they struggle to form normal relationships with us thereafter. In mice, too, isolation in the first few weeks of life results in social impairments; meanwhile, mice isolated for the same length in adulthood show no such effects. This again points to the critical value of social interactions in early life for shaping and guiding brain development.

You may be wondering, where does this development happen in the brain? Do certain regions depend more on social experience to be shaped and matured? Indeed, studies point to a brain area called the *medial prefrontal cortex* (mPFC) as one critical hub where that shaping occurs. Located just behind the center of your forehead, the mPFC plays a central role in social cognition: processing and interpreting social information. It's one of the brain's smarter areas, responsible for some seriously advanced social thinking like *mentalizing*: seeing the world from others' perspectives. For instance, I recently walked into a garden store and quickly caught the name tag of a worker. "Hey, Sandi," I said, "where can I find the lawn bags?" Sandi looked at me quizzically. I could tell she was thinking, *Who the hell is this guy? Do I know him?* Recognizing her confusion, I explained, "I saw your name tag." She gave an understanding smile and confirmed my suspicion: "Ah, thank you. I didn't think we knew each other!" In that moment my brain had mentalized, putting myself in her shoes to perceive that she hadn't seen me read her name tag. If I had been in a brain scanner, it probably would have detected increased activity in my mPFC.

The mPFC is also responsible for *person perception*: the act of forming impressions of others based on our knowledge of them. As Sandi walked me to the lawn bags in Aisle 30, her brain was probably already placing me into certain category buckets. *Outgoing,* given how openly I'd approached her. *Time-efficient,* considering that I'd asked her for help instead of searching the aisles myself. And perhaps a bit *creepy,* considering how I appeared out of nowhere and knew her name. It would now be *her* mPFC lighting up in the brain scanner as she assigned me these traits.

There's another quality that makes the mPFC unique: It develops much later than the rest of the brain. Most parts of the brain stabilize throughout childhood, but the mPFC remains flexible and continues restructuring into our adult lives. Because of that, it's more sensitive to our *experiences* and can be shaped by the cards we are dealt. This makes perfect sense considering the mPFC's role in social cognition. Not everyone is born into an identical social environment, so each brain must be flexible to learn and fit into whichever social culture it exists in.

As we grow up and enter the world, we learn through social experience. We find out during recess that pushing little Johnny down the slide makes him cry and gets us punished; sharing our lunch with the classmate who forgot theirs earns us social admiration; interrupting others while they talk is met with looks of contempt. Through these lived experiences, we build social models of the world. The mPFC seems to contain these models and uses them to guide decisions about what is or isn't acceptable behavior. In support of this, one case report details two adults who suffered damage to the prefrontal cortex in their first sixteen months of life. As adults, both showed "severely impaired" social skills. They displayed little empathy and were prone to lying and stealing—perhaps because they were unable to properly learn social conventions and realize that these actions violate the rights of others. Meanwhile, people who suffered damage to the prefrontal cortex *in adulthood* were much less likely to show these disruptive, inconsiderate behaviors.

These data show how vital early childhood socialization is, suggesting that PFC development in early life may help us learn moral rules. Some of our earliest interpersonal experiences teach us about the conventions of social exchange, establishing rules that become the brain's template for future actions. That's right: The way you (and your mPFC) behave today may be a product of some early interactions you don't even remember . . .

But what happens to a brain without this social shaping? Unfortunately, research shows that children isolated in early life show a smaller PFC. Without those social experiences to exercise and shape the PFC, those synapses may have missed the chance to grow and mature, leaving the region underdeveloped. Similarly, mice that are isolated in early life show underdeveloped synapses in the mPFC in adulthood. The synapses literally look smaller, and when researchers measure their function using a technique called *slice electrophysiology,*[+] they find the synapses are much less active, too. In the case of Genie, perhaps her mPFC was like a slate that forever remained blank, unshaped by the impression of social experience. Just as a piece of clay requires the gentle pressure of a ceramist's hands to become a vase, the mPFC requires social experiences in early life to grow and learn. A piece of clay that spins in solitude will take no such shape.

In a world full of phones, TVs, video games, and—let's be fair—even books, it's a good reminder to expose children to frequent and variable social settings. Looking back on my own childhood, I'm incredibly grateful that I made my way to the more conversational tables. I sympathize with those I left behind at the more reserved tables, for they may have missed out on some critical social experiences. I hope they found social stimulation in other forms, perhaps where they felt more comfortable, like with family. Those early interactions are so valuable, enabling brain development in a way that other experiences can't. And that window of plasticity doesn't stay open forever. We should take advantage of this and do our best to engage children in social activities for the sake of their development.

Don't leave the proverbial clay spinning on the wheel alone.

## VISIT YOUR GRANDPARENTS

The arc of life follows a predictable pattern, at least when you look at who we spend our time with. Research shows that we go through reliable social waves as we age, spending our time with different people at various points in life.

In childhood, we're constantly surrounded by our parents and siblings. Most people will never spend as much time with family as they do during those early years. Then when we reach adolescence, the amount of time spent with friends peaks. Through our late teens and early twenties, we undergo a major social shift characterized by a sharp decline in time spent with family and friends. Sadly, we never return to those peaks. Meanwhile, time spent with coworkers and partners rises dramatically, ushering in a new adult lifestyle. Kids come along next, and our time with them peaks in our thirties—another golden age. Life is good, because we're in the frequent company of our partner and children.

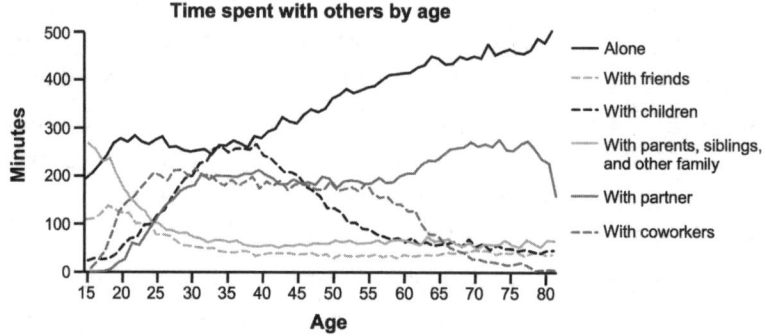

In our forties and fifties, things begin to change. As our kids grow up and move out, we spend less time with them and more time alone. Another major shift toward isolation comes in our mid-sixties as we reach retirement, and time with coworkers bottoms out. In its absence, time spent alone continues to rise. By the time we reach our seventies, the average person is spending over *seven hours a day* alone. Our parents and even some of our friends may no longer be with us. Coworkers are a long-gone afterthought. Even our children don't have as much time for us as they used to. Looking at the data, you can see that isolation comes on gradually. It starts in our forties and progressively builds to a dangerous crest in our eighties.

This is a serious problem. In fact, I would argue that it may be the largest unspoken health crisis of our time. It's an absolute tragedy that the greatest and most threatening peak of isolation coincides directly with old age—a time where we are already more susceptible to health complications like dementia, diabetes, and chronic obstructive pulmonary disease. As Alzheimer's researcher Dr. Joshua Crapser put it, "When you throw isolation in there, it's just boosting all the other health issues." Considering that lonely people are at higher risk of strokes, heart attacks, and other health issues, old age is almost certainly the *worst* and most *vulnerable* time for isolation to strike. Research suggests that in people over sixty-five years old, social isolation increases the risk of death by a whopping 78 percent in men and 57 percent in women. The data don't lie.

Being isolated is also no good if you hope to avoid dementia and Alzheimer's disease. Research shows that isolated seniors are at higher risk of developing dementia, and their memory declines at twice the rate of those who are more socially integrated. Isolation in later life has also been linked to a *thinning* of the cortex (the brain's outer layer) and a shrinking of the hippocampus (the brain's primary memory center). But how could it be, that isolation is literally *shrinking* the brain and increasing the risk of dementia?

The reason could be a familiar one from earlier: dysregulation of the HPA axis. Recall that chronic isolation increases cortisol levels, which can drive harmful effects like inflammation. Well, dementia patients with *higher cortisol levels* show a *more rapid acceleration* of their symptoms over time. Higher cortisol levels have also been associated with faster degeneration of the hippocampus in people with mild cognitive impairment—an early stage of Alzheimer's. So what does this all mean? The way I see it, isolation in later life may be driving up cortisol and inflammation, thereby accelerating the health decline we typically face as we age.

This cannot be ignored. The evidence speaks loudly and clearly—a declaration of the irreplaceable value of social connection. It makes a powerful statement: that older people *need* connection, and moreover, that *all* people should prioritize socializing now—wherever we are in life—because building social habits and routines will pay off tremendously in the long term. Especially considering that brain plasticity becomes more restricted and precious as we age, the sooner we build social habits into our lives and resist the harm of isolation, the better.

Prioritizing interaction can also benefit you for other reasons. Research shows that socializing can offer protection from cognitive decline in later life thanks to something called *cognitive reserve*. Here's how it works:

For a moment, consider how it feels to be face-to-face with another person, engaged in conversation. Your eyes scan them, searching for information embedded within their posture, their facial expression, the appearance of their skin and hair. As they speak, you piece together meaning from their words, factoring in their tone. Was that statement genuine or sarcastic? You try to maintain an appropriate amount of eye contact—not too much, nor too little. You reply in turn, after leaving enough time for them to finish speaking. Did you cut them off? You craft an appropriate response based on your relationship with them—is it professional or casual? As the conversation trails off, you frantically search for a topic to avoid an awkward pause. Another person approaches the conversation to join, and the process begins again—building in the added complexity of their behavior. Do you know each other? Do *they* know each other? Should we start a new topic or just continue? I'm getting tired just thinking about it!

With every interaction, two (or more) brains perform this complex, beautiful song and dance together. And yet we hardly even consider these experiences as a form of heavy lifting for the brain. Social interactions are *complicated*, requiring the brain to judge dozens of streams of information. This requires a great deal of processing power.

Because of this, social interactions are like workouts for the brain. Socializing engages and challenges many brain regions, which helps strengthen and build the corresponding circuitry just as lifting weights can build muscle. As a result, people who engage in more social interactions literally have larger brains and more gray matter. This is tremendously valuable because in *all of our brains*, synapses shrivel and die as we age. We are bound to lose countless synapses as we grow old—it's just part of life. But if we have extra brain volume—thanks to a lifetime of social interaction—it may protect us against the effects of aging and neurodegeneration. This bonus stash of brain tissue may act like a buffer to help us resist cognitive decline, which is why neuroscientists call it *cognitive reserve*. It's like how an army of ten thousand troops can better tolerate the loss of one hundred soldiers than a battalion of one thousand can. Similarly, a larger, more socialized brain can continue functioning at a high level even when certain synapses are pruned and lost. Indeed, a study of more than six thousand seniors found that those with larger social networks and more social engagement demonstrated greater cognitive function and slower cognitive decline as they aged.

Staying socially engaged is a smart bet, and it's never too late to start. I really do think of it like an exercise regimen for the brain. And yes, while those devoted gym rats who've been exercising for years typically fare well through aging, that doesn't mean an eighty-five-year-old who picks up weights for the first time won't see results. Similarly, late-stage social interventions may also be effective. For instance, senior citizens enrolled in a three-to-six-month social activation program showed various health benefits compared to controls, including lower A1c levels (a marker for diabetes) and rises in hormones like testosterone and estradiol. Brain imaging[+] shows that social interventions in later life can strengthen the connections between a group of brain areas called the *dorsal attention network*, which work together to guide attention. To me, this is a great example of how social interaction is like an exercise for the brain. When you talk with someone, you must be quite focused and engaged—much more than when you do other activities like watch TV. It makes perfect sense, then, that having more interactions would strengthen a brain network involved in attention, as you're *flexing* those attentional brain muscles.

Old age is a sensitive and important period, where we should be especially deliberate about how we allocate our time. It is my strong opinion that we should be spending as little as possible alone. By spending time with the elderly, we may literally be extending their life. When Spanish researchers had old mice interact with a younger mouse for just fifteen minutes a day, the older mice lived significantly longer, surviving an average of 120 weeks versus just 90 weeks in unsocialized mice. Let me repeat that for emphasis: The mice lived *33 percent longer* just from interacting with another mouse for *fifteen minutes* per day! If we saw even 1 percent of this effect in humans, would it not be worth it to invest in spending more time with the elderly? In a different study, old mice showed lower levels of inflammation and oxidative stress after living with younger mice for two months. They were also more physically active, were better at walking a tightrope, and lived significantly longer. Simply incredible.

So please, go visit your grandparents, parents, siblings, or friends.

Consider it a favor for their brain.

⟨⟩

When I spoke with Dr. Crapser about isolation, he noted two important points that are absolutely worth mentioning. First, isolation occurs in gradients. "It's a continuum, not a binary scale." In other words, there is no threshold where we suddenly become unhealthily isolated. Rather, the risk of health effects scales with the severity and duration. If you have one good friend that you see once a month, you may not be totally isolated, but you're objectively *more* isolated than someone who has a dozen friends and sees them weekly.

Secondly, there's a major difference between being *isolated*—the objective state of being alone—and being *lonely*—feeling that your social needs are not met. As Dr. Crapser put it, "You can feel lonely in a crowded place or fulfilled in isolation." This is an important distinction. Being by yourself isn't always bad; sometimes it's a relief. If you've ever gone on a long family vacation, you know exactly what I mean. When you get home, you may want nothing more than to sit alone in silence. During this peaceful time, you are *isolated*, but you're not *lonely*. In this context, the solitude is pleasant. On the other hand, if your friend bails on your plans and leaves you to attend a concert alone, you may feel *lonely* even despite being part of a large crowd. This loneliness is not good for your health. Even when factoring in the true amount of time someone spends alone, greater levels of loneliness are associated with increased risk of death. Theoretically, this means that someone could attend a busy concert every single night of their life, but if they never engage with those around them, they may still suffer health consequences.

This is an important excuse to step back and reflect on the current state of *your* social life. A 2018 survey found that 43 percent of Americans felt isolated from others, and only 53 percent have meaningful in-person social interactions daily. Where do you stand on these questions? Are you satisfied with your level of social engagement? Do you feel isolated? How many days per week do you have meaningful interactions? When friends invite you out, how often do you find an excuse to stay home? Do you work in the office every day, or work from home some or all of the time?

Our world is changing in ways that are making us more solitary, and I believe we need to carefully interrogate the effects of those changes in light of the science of isolation. As we go about our lives and make social decisions, we should be aware of the ramifications of *not* interacting. So next time the phone starts to buzz on your desk, think hard before you hit decline.

### Key Takeaways

1. Being isolated is a form of stress, and the brain and body treat it as such. Isolation is associated with depression, anxiety, suicidality, and increased risk of death by any cause.

2. Long-term isolation may lead to chronic inflammation through elevated cortisol levels.

3. Since the brain is highly plastic in early life, socializing in childhood is critical for training social skills and shaping brain areas involved in social cognition. Children should be exposed to ample social interactions to aid their development.

4. As humans age, we tend to spend more and more time alone. This isolation is dangerous; it increases the risk of dementia and death. Living a socially integrated life may stave off dementia due to cognitive reserve. Socializing is of critical importance for the elderly.

5. Social isolation is a continuum, not a binary scale.

6. Loneliness and isolation are distinct, but both are unhealthy.

To view the references cited in this chapter, please visit benrein .com/book.

# II

Nurturing Your
Brain in a Post-
Interaction World

# 3

# BUILDING SOCIAL HABITS

## Working Around the Brain's Natural Limitations

Through the first few chapters, we've laid down some crucial groundwork to wrap your head around the upsides of socializing and the downsides of isolating through the lens of neuroscience and biology. The research shows clearly that our social routines have a powerful hold over our minds, brains, bodies, and health. Part I began to answer the title question of this book: **Why _do_ brains need friends?**

If I did a decent enough job, you may be contemplating your social life in a new way. Perhaps you've reflected on your recent habits and realized there's room to strengthen your social practices. Given the hard truths that **(1) we live in a divided world** and **(2) division is the enemy of brain health**, maybe you're thinking of expanding your social life and building new habits.

I surely hope you are. In fact, I'll just come out and say it directly: _I really think you should spend more time socializing._ This can be with your parents, siblings, friends, coworkers, cousins, neighbors, classmates, or even complete strangers. I have a challenge for you. Next time you're sitting around in a bus, train, taxi, waiting room, restaurant, or anywhere else you might sit and wait, strike up a conversation with someone nearby. It doesn't matter who. Be spontaneous. If there's nobody around, call up a friend or family member. Just see what happens. My guess, based on the findings discussed in chapter 1, is that you'll experience a bump in mood. It's a simple biology hack we can all employ, and one that has always been hiding in plain sight.

Considering the possibility that you'll begin socializing more, it's essential to remember **Hard Truth No. 3: The brain has internal shortcomings that can drive us apart.** It's not just *external* factors like social media that are dividing us, but also our natural social pitfalls. In many social situations, the brain does a frankly shitty job. It makes bad predictions and suffers errors of calculation that can deprive of us of necessary connection. I believe we can be more effective interactors if we are equipped with an awareness of these flaws. In this chapter, you'll learn about some limitations that can hold you back in interactions and ways to maximize your social reward. I hope that by grasping this knowledge, you will be able to build the most effective social habits possible and learn to thrive in a post-interaction world.

## THE BRAIN MAKES BAD PREDICTIONS

I just challenged you to interact with a stranger and made a bold prediction: that it would boost your mood. Now, I don't claim to be a mind reader, but I'm going to make another prediction about what you're thinking. When you read that, I suspect you may have thought something like, *That's a nice idea, but I don't think it's for me.*

If so, I completely get that on a personal level. I've been there, reading a book that offers some prescriptive takeaway and thinking to myself, *What a lovely idea!*, knowing full well that I will absolutely never attempt the suggested activity. But I ask you to consider this differently.

It's normal to expect that you won't feel better after a conversation. Seriously, science shows it's *statistically probable* that you feel this way. This is **Bad Prediction No. 1: We tend to underestimate how much we'll enjoy conversations.**

Remember Dr. Epley's studies from chapter 1, where the subjects had to talk to a random stranger on the train or bus? Well, in describing the studies, I left out an important detail. Before everyone boarded, they were asked to predict how pleasant their commute would be. For some reason, the participants consistently underestimated how much they would enjoy talking with a stranger. When they reached their destination, they were much happier than they'd expected to be. It might seem like a simple conversation couldn't possibly impact your mood, but the results may surprise you.

I think of this often—usually when friends invite me out on a weekend and I'm feeling lazy. In my head, meeting up sometimes feels like a lot of work with little payoff. Meanwhile, lounging on the couch seems *much* more satisfying. But in those moments, I try to think of the people in Epley's studies, waiting doubtingly for a train or bus ride that they're about to unexpectedly enjoy. I remember that social interactions usually exceed our expectations. And when I join my friends, I never regret it. If we let Bad Prediction No. 1 win, we might hold ourselves back from valuable interactions. And unfortunately, this is not the only thing that restrains us. This brings us to **Bad Prediction No. 2: We expect rejection.**

Starting a conversation with a stranger can be intimidating. I mean really, think about being one of the train riders in Dr. Epley's study. Would you feel unsure about starting a conversation with a stranger? They certainly did. In fact, they predicted that *more than half* of the people they approached would reject their attempt at conversation. However, those expectations were completely wrong. The actual outcome was very different: a rejection rate of *0 percent*. Not a single stranger turned down the invitation to talk—and remember, these were real people on real trains.

This actually makes sense to me. If a stranger approached you for a friendly chat, would you turn them down? Assuming they did it in a normal, socially acceptable, and polite way, I certainly wouldn't. In fact, I think I'd be quite happy to pass the time in their company instead of staring impatiently out the window or scrolling mindlessly on my phone. We shouldn't let a fear of rejection hold us back, as these fears are probably overblown.

Now, let's suppose you overcome Bad Predictions 1 and 2, and you convince yourself to interact with someone. Congratulations! However, the brain's miscalculations don't end here. Once we get into a conversation, we fall victim to a whole host of other bad predictions, like **Bad Prediction No. 3: People tend to misjudge the hedonic trajectory of a conversation.** In other words, we wrongly assume that conversations will *get worse* as they go on.

The evidence for this comes from a study in which two hundred strangers were paired up for a chat, creating a hundred conversations. After the pairs spent a few minutes getting to know each other, the interactions were abruptly stopped. Then things got interesting. Half of the conversations ended there: no more chatting allowed. These subjects were asked to predict how much they would have enjoyed the conversation if it had continued for another four sessions, and were accordingly called the *predictors*. Sadly, they had a gloomy outlook on the hypothetical future of their conversation, predicting that it would become less and less enjoyable as it dragged on. Meanwhile, the other half of the participants—termed the *experiencers*—got to see this reality play out. They were allowed to continue talking for another four sessions, each lasting a few minutes, and continuously rated their enjoyment throughout. It turned out those pessimistic predictors were way off. In reality, the experiencers reported a steady level of enjoyment through all four interactions.

While there's certainly an upper limit to how long you can joyfully spend in conversation, misjudging the hedonic trajectory of conversations might lead you to depart from them too early, driven by a false prediction. As a result, you may be depriving yourself of valuable connection.

Finally, we come to **Bad Prediction No. 4: People tend to un-**

**derestimate their conversational skills.** In a revealing 2023 study, people were asked to rank how their abilities compared to others their age. Across the board, people were pretty damn confident. They appraised themselves as being better than most people at reading, doing their job, maintaining their hygiene, and many others. Of the twenty skills they were asked about, the *only* one that people regarded themselves as worse than average at (below the 50 percent mark) was, you guessed it, initiating and sustaining conversation. So if you think you're bad at talking to people, you're not alone.

If (and when) we decide to shake things up and prioritize social connection, we must be aware of these natural pitfalls. While a plethora of external factors may be challenging our ability to connect, we're also fighting internal battles. By recognizing these hang-ups and realizing how they might hold us back, maybe we can more confidently drive ourselves into interactions and reap the greatest benefits.

If you're still unsure, let me throw one final incentive at you. We've established that talking with a stranger might make your day. But it might make *theirs*, too. In Epley's study on the train, the strangers on the receiving end of spontaneous conversations also got pleasure from the experience. Even though they didn't start the conversation themselves, they still enjoyed it. When we leverage interaction to improve our mood, we aren't just doing something good for our own well-being; we're also spreading the wealth to whoever we engage.

In those silent moments in buses or waiting rooms, while many of us are simply passing the time, we may be holding ourselves back unnecessarily. We are intensely social beings, but we suppress these urges due to expectations that are probably wrong. The person across from you may be just as interested as you are to strike up a conversation, and there's only one way to find out.

## SELFISH SELFLESSNESS

Suppose you're on a bus and you overhear someone nearby compliment another rider. They say, "Hey, I really like your sweater." How do you think the recipient of that compliment would feel? Would it improve their mood? Do you think they would feel annoyed, bothered, or uncomfortable?

Now imagine a slightly different scenario: *You* are the one giving that same compliment. You lean toward the person across from you and say the same thing: "Hey, I really like your sweater." How do you think *that* person would feel?

If you felt more doubtful or uncertain about the second scenario, you're not alone. Research shows that people misjudge the impact of *their* compliments but can accurately judge the effects of *others'* compliments. For some reason, we think it's weird to compliment someone but find it normal when other people do so. Sorry, I made it sound like we were done with bad predictions, but this is actually **Bad Prediction No. 5: We *underestimate* how much compliments will positively impact others and *overestimate* how much they will negatively affect them.** What an unfortunate mental pitfall!

The truth is, people really enjoy receiving compliments. Most of the time they're hardly bothered or uncomfortable—unless you make it weird. At the end of the day, a compliment is a compliment, and it makes people feel good. You should try it. People also feel more likely to compliment others after they've done it once, so maybe there's just a threshold of awkwardness we need to overcome. Once we do that, we realize it's not as scary as it sounds.

Oh, and by the way: Complimenting others isn't just productive because it's a nice thing to do—it actually benefits *you.*

Yes, although it may be counterintuitive, it's true. While acts of kindness are typically meant to benefit others, studies show that the people who perform them reap improvements in their happiness and well-being. Moreover, the greater the number of acts of kindness that people perform, the happier they are. It doesn't seem to matter whether the recipient of the kind act is a stranger or a close friend. In both cases, doing a kind act has similar effects on mood. This suggests an unexpected conclusion: In doing something nice for someone, *you* benefit. In other words, **acts of *selflessness* can be done with *selfish* motivations**. I call this *selfish selflessness*.

Gratitude is just the same. Expressing gratitude to others can have positive effects on your well-being, but sadly, people similarly misjudge the consequences of gratitude. They tend to underestimate how good expressing their gratitude will make others feel and overestimate how awkward it will make them feel, and consequently are less willing to express gratitude. It seems that we may be getting in our own way, stumbling over shortcomings in human nature that we're not even aware of.

The other day, I thought of an old friend that I hadn't connected with in a while. I felt remorse about some decisions I'd made in the past, where I had let business get between us. I wanted him to know that although our relationship had hit a few speed bumps, I still appreciated his friendship. So despite my hesitations and fears, I filmed a three-minute video in which I spoke candidly and got those feelings off my chest. I expressed a deep gratitude for the friendship we shared and the ways he has been there for me. Not long after sending it, I received a video in return. "Ben, this made my week," he said. "That video was *everything*." My joy was palpable. It was an exercise in gratitude, and one I would recommend to anyone.

So remember, it's no crime to compliment someone. Expressing gratitude is also a terrific way to deepen your bonds and get a boost. Sometimes the best thing you can do for someone is to simply be nice, and as an added benefit, it may be just as constructive for you as it is for them. Seriously, give it a try. Make it a goal for tomorrow to either give someone a compliment or express your gratitude and see how it makes you feel. You won't regret it.

## INTROVERTS AND EXTROVERTS

I have a (bad) habit of attempting to read minds, and I'm about to make yet another prediction. When you read about the benefits of socializing in chapter 1, did you furrow your brow and think to yourself, *But what about introverts?*

It seems pretty unlikely that socializing would bring an introvert joy; after all, isn't that what makes them an introvert? Introverted people tend to enjoy being alone; they may feel exhausted and overstimulated at parties and social gatherings. In contrast, extroverts take pleasure in these activities. They're energized and invigorated by social gatherings. Introverts tend to be quieter and more reserved, while extroverts are more talkative, outgoing, and socially assertive. But does this mean that introverts don't enjoy interactions at all? This can't be.

Remember the study where the students wore a wire around campus—I mean, the "E.A.R."—and they tracked how conversations influence mood? You may find this surprising, but both introverted and extroverted students showed *equivalent* boosts in happiness after conversations. Introverts even showed bigger effects in some cases—for example, they reported feeling more *socially connected* after deep conversations. This suggests that naturally occurring conversations may be equally beneficial for introverts and extroverts, which makes sense to me. Even for an introvert, the social contact that we *deliberately* seek or select can be pleasant— otherwise we would simply not participate. But how do introverts fare in *forced* interactions, as when Dr. Epley had people approach a stranger on a train?

Interestingly, both introverts and extroverts enjoyed their commute more when they talked with a stranger. Apparently, even introverts can enjoy a forced interaction! However, there was one key difference: The effects were *larger* for extroverts. For the subjects who were naturally more outgoing and talkative, talking with a stranger offered greater benefits. All in all, it seems that while there are some differences between the groups, a *single con-*

*versation* can boost the mood of introverts and extroverts alike. Consistent with this, another study found that when extremely introverted people had to *act* extroverted for a ten-minute group discussion, they surprisingly showed a strong positive surge in their mood. We can therefore summarize the effects of single conversation like this:

| | Introvert | Extrovert |
|---|---|---|
| Organic conversation | 😄 | 😄 |
| Forced conversation | 🙂 | 😄 |

But what happens when we drag out the time scale? For example, what if an introvert has to act like an extrovert *for a whole week*? Is that still helpful, or is it basically torture?

It's basically torture. When shy, reserved people pretend to be extroverted for a week straight, their mood actually *drops*. For the most introverted people of all, a week of extroversion can leave them feeling exhausted and out of character. Meanwhile, when naturally extroverted social butterflies have a hyperinteractive week, they end up feeling much happier and less tired.

It seems that short bursts of sociability can be uplifting even for those who score very low in extroversion. But as time goes on, this process wears on introverts, leaving them in worse shape. It's almost like acting extroverted is a medicine that just one dose of can benefit all people, but introverts suffer more side effects with long-term use.

This suggests that if an introvert and an extrovert go out to lunch together, they will probably both leave in a better mood. But if they go on a weeklong vacation together—staying in the same hotel room and never having time apart—the extrovert will come home refreshed while the introvert will probably be exhausted and irritated. In other words, *trait extroversion*—a stable measurement

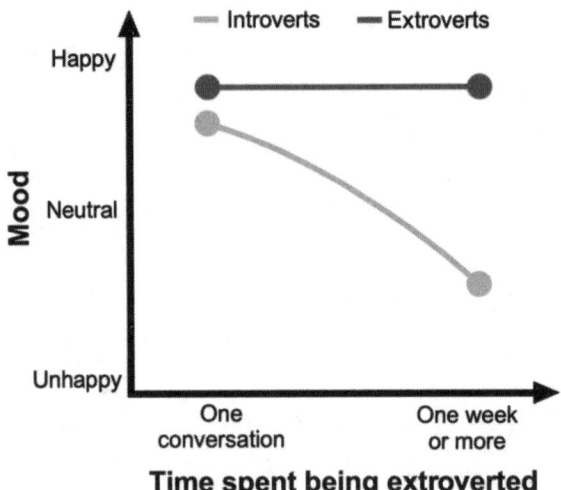

**Time spent being extroverted**

of how sociable someone is—is critical in determining how much they will benefit from social activities. The more extroverted you are, the more likely you are to benefit from extended periods of sociality. Especially before you start building more social habits, it's important to understand what level of socializing brings you joy, including where your upper limit is.

In many ways, extroversion is like the social continuum I imagined in elementary school. Humans can range anywhere from intensely introverted to extravagantly extroverted, and we each fall somewhere on this spectrum. To figure out where *you* stand—and thus how much social engagement is best for you to get to maximize the mood benefits—I've included below a questionnaire for determining extroversion. This scale was adapted slightly from a personality assessment called the Big Five Aspects Scale. Feel free to write your answers directly on the page if you'd like—after all, it's your book. Don't think too hard about it; just follow your gut feelings:

Indicate the number below that feels
appropriate to describe yourself:

| Disagree strongly | 1 |
| Disagree a little | 2 |
| Neither agree nor disagree | 3 |
| Agree a little | 4 |
| Agree strongly | 5 |

1.  I make friends easily.                          _____

2.  I warm up quickly to others.                    _____

3.  I show my feelings when I'm happy.              _____

4.  I have a lot of fun.                             _____

5.  I laugh a lot.                                   _____

6.  I take charge.                                   _____

7.  I have a strong personality.                     _____

8.  I know how to captivate people.                 _____

9.  I see myself as a good leader.                   _____

10. I can talk others into doing things.             _____

11. I am the first to act.                           _____

12. I'm easy to get to know.                         _____

13.   I don't keep others at a distance.          _____

14.   I reveal a lot about myself.                _____

15.   I often get caught up in the excitement.    _____

16.   I am a very enthusiastic person.            _____

17.   I have a talent for influencing people.     _____

18.   I don't wait for others to lead the way.    _____

19.   I don't hold back my opinions.              _____

20.   I have an assertive personality.            _____

**TOTAL**                                         _____

Add up all your values. The maximum achievable score is 100, which would represent the highest levels of extroversion, or the deep right end of our continuum. The lowest possible score is 20, which would represent the lowest levels of extroversion—the far left side of the continuum. There's no real scientific benchmark for what's average, but if you answered neutral (3) for each question, your final score would be 60, which we can consider the center of the social continuum. Based on this, we might classify scores as follows:

20 to 40 = strong introversion

41 to 60 = moderate introversion

61 to 70 = mild extroversion

71 to 80 = moderate extroversion

81 and up = strong extroversion

Keep in mind, the higher your score is, the more likely you may be to benefit from building regular social habits into your life. For those who score lower, you may be better served by sprinkling in occasional interactions at whatever pace is comfortable. Remember, even introverted people can experience mood boosts from single conversations. It's all about finding the right cadence and balance for you. Also, note that interacting with a *diverse* group of people is associated with the greatest effects on well-being, so you may be better served by hanging out with a few different people than by spending all your time with one person.

Before we move on, I want to acknowledge something important buried within this topic. In fact, it's the exact question that first inspired me to study this field—a curiosity I held throughout childhood, and something I wondered about in the waking hours after my life-altering nightmare. **Why do we all land somewhere different on this scale?**

Why is your score different from mine? What's going on in your brain that makes it different from mine? It turns out, these differences in social traits *do* map to differences in the brain. Introverts and extroverts show actual differences in certain brain structures like the *putamen*, a brain area involved in social reward processing. Extroverts show different brain responses to rewards, and some evidence suggests that their dopamine systems function differently (for example, they may carry certain variants of dopamine-related genes). Bottom line: The brains of extroverts and introverts genuinely do *look* and *act* differently, and that's likely the result of everyone's genetics and life experience. We are who we are because of where we come from and where we've been. Your unique combination of *nature and nurture* may determine how much pleasure your brain experiences from being around others, but at the end of the day, everybody needs social connection.

## THE SOCIAL DIET

Considering that we all exist on this social continuum, the social needs and preferences can differ greatly from person to person. I believe this detail may be missing from the public conversation about our "loneliness epidemic." A blanket prescription to socialize is not entirely useful for everyone. There are more nuances on a biological scale to this. Those on the higher end of our continuum—scoring closer to 100—may need more frequent social contact, while those on the lower end may be comfortable with less. The amount of interaction an extrovert prefers could be totally draining for someone more introverted, while an introvert's social habits could leave an extrovert feeling down.

In comparison, think about how your body responds to certain foods. On the extreme level, when you eat something that makes you sick, the brain is damn sure to remember it. Many people have one specific kind of liquor that they simply *can't* drink because the thought alone makes them nauseous. If you have one of those, I can almost guarantee that you once got very sick drinking that liquor. Am I right? When you're poisoned by a substance like that, the brain basically bans it from your body, making an internal note that it's toxic and you should never touch it again. As a result, just the smell or taste of that liquor is enough to trigger a gag response. That's your brain protecting you from eating something it thinks is dangerous.

Through years of experience, you've crafted a custom diet that suits your body's needs. But what about social interaction? This is also a source of nourishment for the brain. Just like the food we eat, social interactions support our health and well-being . . . and you wouldn't feed yourself garbage, would you?

Unfortunately, the brain isn't so good at tailoring our *social diet*—the variety of interactions we each choose to engage in. We don't gag or fart when we interact with people we don't like. Crafting your ideal social diet is a process that may demand active thought and attention. There are certain reasons why an interaction might leave you feeling worse off, and identifying those important influences may help you leverage interactions to your benefit. Crafting your social diet means learning to recognize what represents pleasant and unpleasant social experiences for you, instead of acting blindly. It's like figuring out which foods make you gassy—but for your social life.

For this, I recommend a simple practice I call *social journaling*, an introspective exercise in which you document how interactions impact you. At the back of this book, I've shared a template that you can use for this, which is also available for free on my website (benrein.com/book). This is intended to help you digest your experiences (pun intended) in an attempt to shape your ideal social diet. By journaling after social events, you can grasp how different features of interactions influence your experience. The goal is ultimately to fine-tune your social life to maximize your enjoyment in social settings. Maybe you hate yelling over loud music in crowded bars, or you feel uncomfortable in one-on-one settings. By tracking your interactions over the course of a week or a month, you will likely notice trends. Work to identify the factors that seem to keep coming up in bad interactions, and those that recur in good ones. Of course you are also welcome to customize this journaling exercise however you wish, adding or removing prompts as you see fit.

On our journey to build connection—in a larger quest to increase our sense of well-being—we must understand why it can be hard to take the medicine we need: time spent together. Our brains, as skilled as they are, have some unhelpful social tendencies that can get in the way of the prosperity we seek, from underestimating the value of interactions to misjudging our own social skills and needs. By knowing *why* we sometimes struggle to prioritize connection, we can better tackle *how* to engage more—and in ways that fit our unique preferences. In the remainder of this book, we will continue to answer both the *why* and the *how*, exploring central questions like how virtual interactions compare to those carried out in person, how drugs can influence and tinker with our social brains, how to become a better interactor, and even how bonding with dogs can support our brain health. But first, we must take an in-depth look at an indispensable element of social interaction that shapes our bonds: *empathy*.

### Key Takeaways

1. Humans are prone to social miscalculations that can jeopardize our ability to form meaningful connections. We underestimate the benefits of an interaction, falsely predict rejection, miscalculate how long to stay in conversation, undervalue our conversational skills, and misjudge the impact of compliments and gratitude.

2. Acts of kindness and generosity can pay off with benefits for the actor. Thus selfish selflessness is the act of being kind not only for someone else's good but also your own.

3. Introverts and extroverts alike benefit from interactions; however, socializing over longer periods can be fatiguing for introverts and result in more side effects.

4. Your level of trait extroversion can indicate how much you'll gain from building social habits.

5. Social journaling can help you build your social diet into a tailored inventory of social experiences that supports your well-being.

To view the references cited in this chapter, please visit benrein .com/book.

# 4

## EMPATHY AND APATHY

### A Tale of Perspective

I was only seventeen years old when I stepped foot on the campus of West Virginia University as an anxious, ambitious, and naive college freshman. Still following the scent of my childhood interest in that social continuum, I was hunting for a chance to work in one of the research labs on campus. I wasn't exactly sure what to expect from working in a lab, and maybe neither do you. But wherever you think this story is going, you're probably wrong . . . unless you're expecting that it ends with me scaring innocent people with a dental drill.

In my first semester, I met a psychology professor named Dr. Daniel McNeil, who turned out to be an important mentor for me. Gray hair, round glasses, blue oxford shirt with a yellow tie, rushing to his office with a coffee-stained mug in one hand and an overstuffed folder in the other, Dr. McNeil is a true old-fashioned academic scholar. His lab studies dental phobia—yes, the fear of the dentist. When I met Dr. McNeil, I realized I could volunteer in *his* lab to learn more about research. Luckily, he was eager and graciously willing to mentor me.

I'll admit, dental phobia wasn't a topic I was particularly interested in (sorry, Dr. McNeil), but I was willing to learn. In truth, I joined the lab mostly to find out if science might be a good career fit for me. Little did I know, I was about to learn a life-altering lesson about a very different topic: *empathy*.

Dr. McNeil's lab was testing a new way to treat dental phobia through *exposure therapy*. The researchers asked people who were afraid of the dentist to spend eleven minutes a day watching a video of a routine dental visit. For those with a strong phobia, being repeatedly exposed to their fear in a safe environment can be helpful for overcoming it. Just think about it—if you completely avoid what you're afraid of, the only way you'll interact with it is through the horror-filled encounters that play out in your mind. On the other hand, if you carefully configure pleasant or neutral real-life experiences around the phobia, you might teach your brain that it's not so scary. The lab hoped that watching this non-threatening dentist visit might help the subjects mitigate their fears.

When it came to measuring someone's fear of the dentist, Dr. McNeil was not messing around. He had turned a whole room in his lab into a fake (but very convincing) dentist's office, complete with a dental chair and all the associated tools—but in the middle of a carpeted room in an old academic building. For those with dental phobia, this was pretty much hell. Each participant in the study was brought in for a fake dental visit, during which they were hooked up to sensors measuring things like their heart rate and their skin conductance—the level of sweat on their skin. If they were afraid, the lab's sensors would detect it.

One day while I was helping in the lab, something happened that I'll never forget. A subject was coming in for their fake dental exam, and the lab needed someone to play the dentist. I happened to be there. Guess who they picked?

As I think back on this memory, it's actually hilarious. Imagine a nervous teen boy in a baggy white coat with a sloppy handful of supplies strolling into your dentist's office and announcing himself as your doctor. I was so nervous. Even though the "patient" knew it wasn't a real dental exam, it still mattered to me. I did *not* want to screw up the study.

I walked in and introduced myself as the dentist, then asked a few questions about their oral health, reading off a clipboard. I felt silly and fumbled some of my lines, but luckily they seemed to take me seriously. Then I peeled a wooden tongue depressor from its white paper wrapping and asked them to "say ahhh." I pressed it softly onto their tongue, counting to ten in my head as I watched the saliva pool around the indent. I tried not to think about how miserable they were, but I could feel the tension rising in the room as their discomfort swelled. Meanwhile, signals representing the patient's rising heartbeat and skin conductance were broadcasting to the next room over, where a team of researchers sat and listened, watching the data stream in.

I withdrew the tongue depressor and nervously tossed it into a small trash bin at my feet. Unfortunately, I knew the patient's dread was about to reach an agonizing peak. To my left, there was a dental drill sitting on a small table. I reached for it cautiously, slowly lifting it between us until it reached eye level. I held it there for a few seconds. Then I pulled the trigger. The drill whirred loudly, filling the small room with its awful mechanical screech. The metrics flared on the computer screen next door as the patient's heart rate climbed precipitously and their skin glossed with sweat. This was more than enough to drive them to near panic. I continued to hold the wailing drill between us, watching the subject's eyes fill with alarm and desperation. Suddenly I felt a jolt of dread slash into me, a chilling tingle rising from the depths of my stomach up into my chest. I wasn't afraid of the dentist. I wasn't the research subject. Yet here I was, forgetting this was a lab experiment and I was the experimenter. My heart rate was soaring, just like my patient's. I felt their terror surge through me, as if it were somehow contagious.

Then I released the trigger, and the drill slowed to a halt. I set it down shakily and quietly exited the room, having totally forgotten about my performance nerves. I was too focused on this new feeling of panic, which I seemed to have caught like a virus. I gathered my books and ventured off to my next class, carrying with me a sense of unease that I would only gradually shake over the next hour. It would be the last time I volunteered to be the dentist in the lab.

## EMOTIONAL CONTAGION

I didn't know it then, but what I'd experienced that day was something called *emotional contagion*. Yes, emotions can be contagious—passing rapidly from person to person—and I had somehow "caught" the subject's panic. Beneath my awareness, my brain was detecting signs of alarm and leading *me* to feel the same intense fear.

But how does this even happen? Can emotions magically jump from one person to another? That doesn't seem possible. Or is it?

Of course not. Human biology isn't magic, though it may seem like it. Emotional contagion can be explained by science, and it turns out that one of the key systems driving it is *facial mimicry*. This is exactly what it sounds like: When you see other people make facial expressions, your face naturally mimics them. Research shows that when someone looks at a happy facial expression, their zygomatic major muscle (which lifts the corners of the mouth) is likely to become slightly activated, gently imitating the action of a smile. Meanwhile, looking at negative facial expressions like anger will engage a muscle above their eyes called the *corrugator supercilii*, lowering the brows into a frown. This imitation happens automatically, and largely beneath our awareness. It's fundamental to the way we interact.

This might be hard to believe; I mean, how did you never notice this before? The truth is, you probably have: If you've ever been a victim of the "contagious yawn" then you've consciously experienced facial mimicry. That's right; empathy researchers have proposed that the reason we catch others' yawns is because our facial muscles automatically imitate the motion of widening the jaw.

But why does this happen in the first place? What do we gain from imitating others' facial expressions?

It turns out, matching someone's facial expression helps us understand their emotions. While you may think information flows only from the brain to the face (i.e., the face smiles because the brain feels happy), it also travels in the other direction. Meaningful signals can feed back from the nerves in the face to help the brain estimate your mood (i.e., the brain is happy because the face is smiling). Maybe you've heard of the classic study where people watched cartoons while holding a pen between their teeth, which forced their face into an artificial smile. Astoundingly, these people rated the cartoons as funnier, suggesting that the grin on their face persuaded their brain to think they were enjoying the cartoons more. Other studies, too, have shown that manipulating someone's face into a smile or frown can propel their mood in that same direction.

This is how facial mimicry drives emotional contagion. When you automatically mimic a friend's facial expression, you begin to feel what they feel. The feedback from your face makes your brain think you're experiencing the same mood, and next thing you know, voilà, you've absorbed their emotions. Emotional contagion thus happens in three steps: *mimicry, feedback,* and *contagion.*

As it happens, this process is surprisingly vital for helping us sense people's emotions. When we can't copy someone's face, we find it harder to understand them. For instance, when highly empathic people held a chopstick between their lips to restrict them from engaging certain facial muscles (try it with your finger or a pencil), they took longer to figure out what emotions other people were expressing. Similarly, people who receive Botox injections (which paralyze the muscles in the face) struggle to recognize emotions from facial expressions. Blocking facial mimicry prevents us from understanding one another, highlighting its importance for empathy. Some evidence suggests that mirror neurons[+]

may be involved, but the exact mechanisms are not entirely sorted out.

What does this all tell us? First, that our bodies are stocked with powerful and sneaky systems that guide our emotions beneath our awareness. The muscles in your face do more than help you express emotions. They also help you *understand* and *take on* others' emotions. This may explain why facial expressions are so important in conversation: They allow us to *feel* others' stories by helping us embody the emotions.

Second, this reveals something major about the human experience that is clearly demonstrated by my dentist story: The people around you can have a powerful and *subliminal* effect on your mood. When emotions arise in the brain next door, they just might penetrate your mind and change your perceptions without you knowing. Sometimes when you look inside yourself and try to figure out where your current mood is coming from, you may find that the emotions aren't yours at all.

And third, we may want to think twice before overdoing the Botox.

Looking back on my experience as the fake dentist, I'm confident that facial mimicry played a big part. As I took in their fear, my face was likely contorting into a similar expression, sowing the seeds for my panic. As the signals fed back from my face to my brain, my emotions followed suit. In retrospect, I probably wasn't the best choice for playing the dentist role.

But emotional contagion isn't just about facial expressions. Our bodies also imitate other things like posture and vocal tone. Just think about this. How would you feel talking with someone who is seated comfortably on a couch, speaking slowly and quietly? In contrast, imagine talking to someone standing rigidly and shrieking in alarm. The differences between the two physical reactions may come down to mimicry, as we embody their expressions and start to feel the same way they do.

The same might happen if you were to watch someone doing something dangerous. For instance, one study spectated the *spec-*

*tators* at a firewalking ritual to see if they experienced emotional contagion, and they sure did. Every June on the summer solstice, the residents of a small Spanish village cross hot coals barefoot while a crowd surrounds them, cheering and playing music. When the firewalkers were strapped into heart monitors, it was unsurprisingly found that their heart rates surged as they crossed the hot coals. But what about those in the crowd? Amazingly, the *relatives'* heart rates tended to rise in tandem, establishing a sort of synchrony with the firewalkers'. Despite the fact that they were simply standing there watching—a very different action from walking on hot coals—their hearts began to race. It was once again a demonstration of emotional contagion.

By the way, if you've been wondering about the results of our study, the dental exposure therapy worked. The people who watched the eleven-minute video every day felt less anxious in the mock dental exam and showed lower heart rates. I learned a lot from my experience in the lab, mostly about doing research and designing studies. I also learned that I probably didn't want to be a dentist. But most important, I learned about how the emotions of others can impact us. As I walked tensely to my next class, I had no doubt that emotions were contagious. A curiosity had been instilled within me. It was one of those moments where I'd wished I had been studying neuroscience instead of psychology. I was left with a profound sense of awe, and I couldn't help wondering about what was happening in the *brain*. It wouldn't be long until I experienced the nightmare that cemented my transition into neuroscience.

## WHAT IS EMPATHY?

*Empathy* has become sort of a buzzword nowadays, and I actually think that's great. I'm *glad* people care so much, because empathy is fundamental to our relationships. Without it, I don't think humans would be here today (and you'll soon understand why). But what exactly *is* empathy?

We can define empathy as "adopting a state that's more appropriate for someone else's situation than your own." In other words, empathy is when you understand or take on another person's emotions. Sounds like emotional contagion, right? Well, there's a key difference. Emotional contagion happens *subconsciously*, while empathy is something that you're aware of. In fact, this is one of the key factors that some researchers use to define empathy. In the case of my catastrophic dentist incident, I didn't know where my nerves had come from—a feature that classifies it as emotional contagion. If I had been aware that my feelings were being driven by the subject in the dental chair, it would have been a case of empathy.

The other day I saw someone post the question online: "Is empathy a strength or a weakness?" That's an easy one: It's a tremendous strength. I mentioned earlier that without empathy, humans might not be here today. I really do believe that. Empathy lets us understand and share someone's emotions *without speaking a single word*. It helps us learn from their experiences without taking part ourselves. It's literally *mind reading*, and it comes standard in most human brains. Empathy is therefore one of the core reasons *why* humans survive better in groups. Perhaps most important, when you take on someone else's pain, you become more motivated to step in and help them, because *you* feel bad. Sure, empathy can hurt, but what would we be without it? If we instead evolved to feel nothing for one another, humanity would be a tragedy.

Because of these many benefits for working in groups, empathy was favored by evolution. Without even thinking about it, we can immediately understand what others are going through, and that can literally save lives. For example, imagine you're a Neanderthal walking through a dense jungle with a handful of friends. Suddenly someone lets out a shriek. You turn to see their face twisted into a grimace as they reach for their foot. Almost immediately, you realize what's happened: They've stepped on something painful. At once, you stop and check the ground around you for thorns or other hazards. Then you rush to help your fallen companion.

This is the value of empathy. By automatically sensing your comrade's misfortune, you increase the chances of your own survival. If you hadn't recognized their pain, you might have thoughtlessly walked over to them and met the same fate. Your empathy also drove you to help them, another major benefit. Most impressive, it took a matter of milliseconds for you to gather this information, and not a word was spoken.

Empathy comes in two flavors: *cognitive empathy* and *emotional empathy*. In many cases, the two swirl together like a delicious cone of chocolate and vanilla soft-serve ice cream. When we experience empathy, we can typically break it down into these two components and isolate the separate experiences, like taking a lick from the chocolate or vanilla ice cream alone.

Cognitive empathy involves *understanding* someone else's thoughts and emotions. It's the ability to wrap your head around what someone is experiencing and feeling based on their social cues. In our jungle example, cognitive empathy allowed you to comprehend that your companion was feeling pain in their foot. You saw their behavior (loud scream, painful face, reaching for their foot) and recognized what was happening in their mind (ouch, my foot!).

Emotional empathy, on the other hand, is the process of *feeling* someone else's emotions. Seeing your pain-stricken friend moaning in agony, you might experience a sense of secondhand discomfort. This would be emotional empathy, as you're partially taking on their emotional experience. You're not feeling actual pain, but you feel a disturbing reaction at the sight of them wailing in pain, holding a bloody, mangled foot.

These two forms of empathy are distinct. Cognitive empathy involves only thoughts (no emotions), while emotional empathy involves only emotions (no thoughts). However, they often exist together. This is why I like to think of them like a soft-serve swirl. They're great on their own, but they really work best together. Typically, one leads into the other. If we feel someone's emotions first (emotional empathy), then we can instantly understand what they're thinking (cognitive empathy). If we first understand what they're thinking (cognitive empathy), then we might be able to feel our way into their emotions (emotional empathy).

A common misconception is that empathy applies only to negative emotions like pain, sadness, or embarrassment. However, empathy also happens for positive emotions. Imagine you're in the crowd at an elementary school graduation while the principal is announcing awards. She tells the story of Liam, a student who went from failing all his classes last year to graduating in the top 5 percent of his class. Enthusiastically she announces that Liam has received the Student of the Year award. He leaps onstage to receive his plaque, and he's beaming with pride. Seeing his pure excitement and joy, you are overwhelmed by a sense of happiness. You break out into a big smile and clap loudly, even tearing up. In this moment, you're stepping into Liam's experience and feeling some of his positive emotions *for him*. This is empathy, too.

I find it incredibly useful to recognize the differences between cognitive and emotional empathy. Once you wrap your head around these principles, you will notice differences in your interactions. You will pick out moments where you experience just one form of empathy or the other, or both together. You will recognize emotional nuances within your interactions that have always been there but never reached your awareness. Combining this with the knowledge of emotional contagion, you will be well equipped to understand your emotions in social settings better than most can. If you ask me, these are great superpowers. I hope you put them to good use.

## WHERE EMPATHY COMES FROM, AND WHERE IT GOES

Although empathy may be essential for our survival, it's surprisingly not something we're born with; *it's something we learn.* Amazingly, this process begins in infancy. Research suggests that emotional mirroring (when parents mimic their baby's facial expressions) is an important early step. Infants with depressed mothers, who are less likely to engage in this mirroring, show lower levels of empathy. Our empathy continues to develop as we progress through childhood, and our lived experiences teach us about how others express emotions—another persuasive reason why socializing in early life is so important. If our parents are good at sensing and responding to our emotions, we learn the value of empathy through this firsthand experience. A parent's sensitivity and responsiveness to their child's emotional cues are referred to as *synchrony*, and research shows that greater parental synchrony in early life is associated with stronger empathic abilities in adolescence. This is a good excuse to practice attentive parenting, noticing and responding to the emotions of your child.

This may leave you wondering if empathy is hard-set once you reach adulthood. Are you stuck forever with whatever bit of empathy you were able to muster from your childhood? Luckily, no—empathy remains flexible. In fact, studies have shown that specific trainings can not only improve empathic abilities but also drive structural changes in the brain! After going through a nine-month program involving meditation and other trainings[+] to enhance empathy, perspective-taking, and compassion (which are described in the appendix in case you're interested in practicing them), subjects showed significant[+] improvements in compassion and perspective-taking. Moreover, these trainings drove synaptic plasticity in empathy-related brain areas like the insula (which you'll learn more about soon) and even caused certain parts of the frontal cortex and temporal cortex to *grow thicker*. If you (like me) are someone who wants for a more empathic world, this is terrific and hopeful news.

However, the experiences we have in adulthood can also send our empathy toppling in the opposite direction. For instance, post-traumatic stress disorder (PTSD) is associated with a reduction in empathy, and unfortunately, so is becoming a doctor. Students tend to become less empathetic as they progress through medical school, which is likely an adaptive response to protect their well-being. With constant exposure to pain and trauma, their brain may intentionally desensitize them to others' suffering and limit them from taking on others' emotions, resulting in a gradual decrease in empathy. Without that shift, the profession may be too emotionally taxing.

As we go through life, our cognitive and emotional empathies also tend to change in specific ways. For most people, emotional empathy grows stronger as we get older. Perhaps this is a sign of wisdom: As we experience more life, we gain a deeper appreciation for others' perspectives and grow more likely to take on their emotions. However, cognitive empathy shows a different pattern throughout life: an inverted U-shape. Performance in cognitive empathy tasks (like the Reading the Mind in the Eyes Test, where you guess someone's emotions from seeing only their eyes) rises through the early twenties, peaks around thirty-five or forty, and then progressively declines. Interestingly, people with a college education score higher on cognitive empathy measures and show less decline in cognitive empathy through old age.

While it may sound like empathy is based only on our experiences—all nurture and no nature—genes can and do factor in. People with certain versions of the oxytocin receptor gene or the serotonin transporter gene show more empathy than others. Meanwhile, other genetic changes can *reduce* empathy. For example, people with variants of a gene called *MAOA* (which breaks down neurotransmitters like serotonin and dopamine) are at higher risk of psychopathy, a condition characterized by reduced empathy. These various genetic changes are probably altering the brain systems behind empathy, either boosting them or holding them back.

I may have come to the realization in my childhood cafeteria that everyone exists somewhere on a spectrum of sociability. Since then, I've come to the realization that *all* traits exist on a spectrum. If you collect a group of fifty or one hundred people, you'll find tons of variability in things like intelligence and athlet-

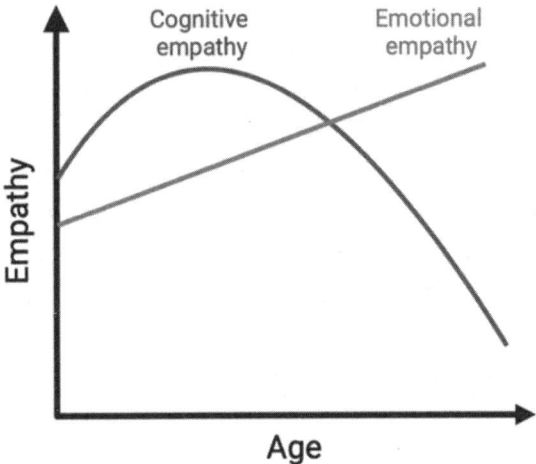

icism. We each exist somewhere on a slider from "extremely low" to "extremely high" levels of these traits, and that position is determined by our genetic background and experience. I believe this is also true for empathy. We all fall somewhere on a spectrum ranging from *no empathy* to *so much goddamn empathy I can't take it,* and there are assuredly even more spectrums *within* this spectrum. For instance, there's a spectrum representing how much empathy you feel for your parents, your best friend, people of different racial backgrounds, those of different religious faiths, various ages, and so on. It all depends on how your brain's empathy systems have been shaped by your genetics and experience. Once again, we are who we are because of where we come from and where we've been.

Speaking of your brain's empathy systems, I'm sure you're wondering how the hell empathy works. What happens in your brain when you *feel someone else's emotions*? If you ask me, this is one of the most fascinating topics in all of science. And more important, it reveals yet another of the brain's natural social pitfalls—a key barrier to empathy that lies deep within our neurochemistry. Let's dive in.

## I FEEL YOUR PAIN

For a moment, I want to revisit the scene from earlier, when you're walking through the jungle with your Neanderthal friends. This time, I want you to *really* imagine yourself there. It's a tropical jungle—hot and humid—and you can feel the moisture in the air. Tall, leafy trees rise above you. There is no trail to follow. You're walking through thick, dense brush, and you can hardly see your feet through the undergrowth. You're with a group of people, and everyone is barefoot. A few minutes ago, you noticed a patch of colorful bushes with thick heavy leaves and the largest thorns you've ever seen on a plant: roughly the size of a finger, with a tip as sharp as a cactus spine.

Who are you with? I want you to assemble the group. It can be some members of your family or a group of close friends. Choose three to four people in total. Put yourself in this jungle with that group of people. Now pick one of them.

The person you've chosen steps on a thorn. A horrible shriek of agony escapes from them as their face twists into a grotesque display of pain and alarm. They jerk their foot from the ground, revealing that a thorn has plunged all the way up through their foot, coming out the top. Deep red blood pools around the edges of the wound. A spray of droplets hits the forest floor beneath them as they drop to the ground weeping, reaching gingerly toward the thorn. I know this may be uncomfortable, but really try to picture yourself in this moment. Imagine it as if it's real.

How do you feel?

My guess is that this paragraph was hard to get through. Maybe you felt a sense of disgust or revulsion, wanting to shrink away from the thought. If you're like me, you may have even felt a sort of sensitivity in one or both feet, imagining the sensation of the thorn piercing through. If you experienced this revolting sensation, I'm very sorry—but I did that on purpose. In fact, that terrible feeling is exactly the point I'm trying to make. That feeling is empathy.

Empathy puts us into someone else's shoes and lets us feel their emotions, and in this case, those emotions are terribly bad. Your comrade's painful experience is reflecting into you through this feeling of disgust. Why? Again, it comes back to evolution. That icky feeling helps you learn two vital things: (1) *Do not* step on a thorn because it sucks enormously, and (2) this person could really use some help. Both of these realizations promote your survival and the success of your group.

Through decades of study, researchers have uncovered the neuroscience beneath this phenomenon, and it's quite remarkable. It turns out that when we see someone in pain, our brains activate in a way that *resembles pain itself.* This is presumably why reading that paragraph felt so uncomfortable: because it engaged some of the brain areas that tell you something hurts.

The evidence for this comes from a paper published by three prolific empathy researchers: Dr. Claus Lamm, Dr. Jean Decety, and Dr. Tania Singer. The trio examined what happens in someone's brain when they look at pictures of painful experiences, like a hand being poked with a needle or a finger caught in a pair of scissors. The images weren't gory; they were just meant to represent painful situations. They found that across a number of studies, two key brain areas were consistently activated: the *cingulate cortex* and the *insular cortex.*

This made perfect sense, as both regions contain a unique type of brain cell called *von Economo neurons,* which are thought to play a special role in empathy and social behavior. When people suffer from degeneration of von Economo neurons, they show a loss of empathy and social awareness. So Lamm, Decety, and Singer reasoned that the cingulate cortex and insular cortex could be some of the brain's core machinery driving empathy for pain. However, they realized something deeper was happening here. Those two areas aren't just random bits of brain tissue. Rather, they're some of the main systems that are directly activated by *pain.* Somehow, observing someone else's pain was enough to activate some of the brain areas that process pain itself! This suggests that empathy for pain truly involves a partial sharing of the experience, which could explain that "please get me away from this *now*" feeling.

Now, there's not a complete 100 percent overlap in the brain between *real pain* and *empathy for pain*. Real pain involves several brain areas that don't come online during empathy, and that's probably a good thing. After all, if we felt the *full extent* of others' pain, it would be unhelpful and probably counterproductive in the end. Being around someone in pain would be too unpleasant to offer help. Nobody would want to be a doctor or nurse—it would be too painful. Football and hockey leagues all over the world would go bankrupt, as nobody would pay money to sit in the stands and feel the full sensation of every collision. Ordering a boxing match or UFC fight on pay-per-view would be absurdly masochistic. And I can assure you, 0 percent of husbands would stay in the hospital room while their wife gave birth. The whole fabric of society would probably fall apart. Yes, that's right, I believe that humanity would crumble if we had *no* empathy or if we had *too much* empathy. We seem to be in the Goldilocks zone: right where we need to be.

Instead, the brain represents only the unpleasant *emotional* piece of pain. If you think about it, physical pain comes with an emotion that might be described as "suffering" or "unpleasant" or "holy shit, please make this stop." This portion of the pain experience has been linked to two brain areas—the cingulate and insula. Coincidentally, those are the same two regions that turn on when we *observe* pain. That means the cringey sense of discomfort you felt when reading that terrible paragraph might have been the feeling of your cingulate and insula activating.

If seeing someone get hurt really *does* activate some of the brain's pain systems, then wouldn't taking painkillers block that empathy? This is the clever question that a group of researchers asked in a 2016 study, where they had people take either a standard 1,000 mg dose of acetaminophen or a placebo before reading about a painful experience (like someone slamming their fingers in a car door). Amazingly, those who took acetaminophen *felt less distressed* while reading these stories, suggesting that empathizing with someone's pain really *does* engage your brain's own pain systems, and blocking those systems limits that empathy. Meanwhile, when asked to rate how painful the experience was for the character in the story, they guessed the same as the placebo group. What does this mean? Well, it suggests that painkillers don't block *cognitive* empathy; they only reduce how much distress we feel in response. In other words, if you popped some acetaminophen before reading this chapter, you would have felt *less* uncomfortable reading about the thorn incident, but you would still understand just how painful it was for your companion.

Interestingly, the same seems to be true for painkillers and *social* pain (if you've ever heard someone talking trash about you or been picked last in gym class, *that's* what I mean by social pain). This evidence comes from an experiment where subjects played a computer game called Cyberball, which involves tossing around a virtual ball with other players. Cyberball is commonly used in neuroscience studies because it can be played on a screen while lying in a brain scanner, allowing researchers to study what the brain does during a social exchange (it would be quite difficult to play catch in the tube of an MRI machine).

Anyhow, as the subjects were playing virtual catch, one player slowly stopped getting the ball thrown to them—a rather embarrassing event. The other players continued tossing it around, but the excluded player stood by waiting for a pass that never came. As you may know from experience, being socially excluded can be painful and humiliating. Later, the other players were asked how they felt watching this happen, and most people reported feeling very uncomfortable. However, those who took acetaminophen reported much less distress. This makes sense, considering that social and physical pain also show significant overlap in the brain (more on that later).

Overall, the neuroscience of empathy reflects exactly what it feels like: that we actually do take on people's emotions by stepping into a similar pattern of brain activity. But there's one problem: The brain doesn't always use this ability.

We feel more empathy in some circumstances and less in others. Sometimes we may not feel any empathy at all. Why is that? Well, this brings us back to **Hard Truth No. 3: The brain has internal shortcomings that can drive us apart**. As you're about to learn, the brain can be rather *selective* with its empathy, reserving it for certain people. If we aspire to be a civilization built on caring for one another, we must grasp the influences that boost our brain's empathy systems or take them offline. You may be surprised to learn that in many situations, you're less in control than you think.

## NOW YOU FEEL IT, NOW YOU DON'T

It turns out, the brain can be kind of a petty organ. Empathy isn't as steady and reliable as we might hope. Rather, it tends to be very context dependent, and our feelings toward someone can differ based on the situation. It depends on a bunch of factors.

Some of these factors are obvious, like how strong the other person's emotions are. You will probably feel more empathy for someone sobbing intensely as opposed to someone frowning slightly. The more intense their experience is, the more intense yours will be. People also tend to feel more empathy when they've been in the demonstrator's situation before. For instance, if you've slammed your fingers in a car door, you might wince a bit harder at the thought of it happening to someone else. In contrast, men may struggle to comprehend what it's like to go through childbirth because they've never experienced it themselves. As an extreme example of this, people born with the total inability to experience pain (a very rare condition called *congenital insensitivity to pain*) tend to misjudge others' pain, underestimating the severity.

Other factors that influence empathy are more nuanced, like

the context of the situation. If your friend told you that they were stabbed with a hundred needles earlier today, you would probably feel tremendous empathy for them. But if they then explained that it was acupuncture and it made them feel much better, you would no longer empathize. Context matters.

Similarly, you might not empathize as much with someone who has treated you unfairly. Dr. Tania Singer—who teamed up with Lamm and Decety earlier in the chapter—had subjects play a financial game with someone who was secretly an actor. Sometimes the actor was cooperative in the game, while other times the actor totally screwed over the subject and took all their money. Later the subjects were put in a brain scanner while they watched this actor receive painful shocks on their hands (I know, hardcore stuff). Remarkably, people showed *less activity* in empathy-related brain areas if the actor had treated them unfairly. Meanwhile, men (but not women) showed activity in *reward*-related brain areas—suggesting that they took some joy in this revenge. Moral of the story: Your brain takes note of how others treat you, and withholds its empathy accordingly (and men might be a bit more twisted than women).

As you are beginning to see, the brain cares a lot about *whom* you're empathizing with. You might feel extreme empathy for one person and hardly any for someone else. A few pages back, I asked you to picture a specific group of people with you in the jungle: either close friends or family. Sorry to say, I also did that on purpose . . . because I really wanted you to feel their pain.

People tend to feel more empathy for those they're closer with, and research shows that we can track these differences back to brain activity. When people watch their best friend get excluded from the passing rotation in that Cyberball game, they show increased activity in those brain areas from earlier that drive the emotional piece of pain: the anterior cingulate cortex and insular cortex. But when they watch a *stranger* get excluded, different brain areas light up, like the prefrontal cortex, which is more involved in cognitive empathy. This suggests that when we watch our friends suffer, we may be more likely to *feel* their pain; but when it's a stranger, the brain is more focused on *understanding* that person's experience, with less of an emotional attachment.

In many ways, this seems intuitive. I mean, duh, of course you care more about your best friend's feelings than a stranger's. But there's something else buried beneath this seemingly obvious take-away: The brain's empathy systems favor people who are *like us*. When people view their friend as more similar to them and report a higher level of *self-other overlap*, they show stronger empathic responses in the brain.

Self-other overlap is like a Venn diagram between you and another person. Think of someone in your life (maybe start with a parent or sibling) and imagine yourself as one circle in a Venn diagram, and them as the other circle. How much do your circles overlap? Consider your personality, educational background, interests, race, profession, political ideology, sense of humor, religion, sexuality, *everything*. If you're identical in every way, the circles might overlap completely. But if you're total opposites, you may share only a tiny sliver. Self-other overlap can be used as a simple measuring stick for how likely you are to empathize with someone: the greater the overlap, the more likely your brain is to engage its empathy circuits for this person. In theory, this can be applied to anyone—a stranger in the car behind you driving recklessly, a person on the street asking for change, or a character in a TV show. In just a moment of thought, you can imagine this Venn diagram to better understand your brain's empathy systems.

Why is it that self-other overlap has such a strong bearing on empathy? Think about this. In ancient civilization, it would've been helpful to experience stronger empathy for those within your tribe and less for those in other tribes. If your comrade is struggling in battle with a predator, it's helpful for you to feel naturally motivated to jump in and help them because it promotes the survival of your group. By taking on their pain and despair, you become more likely to intervene. However, if it's a member of an

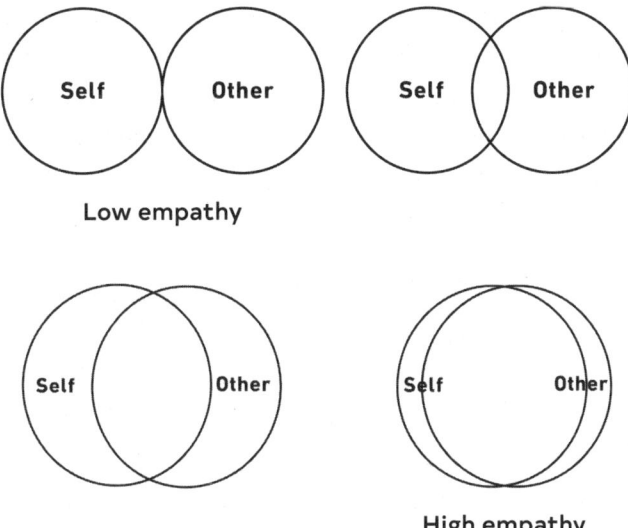

Low empathy

High empathy

opposing clan, you're better off leaving them to die. It's actually advantageous for you to feel no emotions while watching them perish, as it prevents you from risking your life to help someone who will only stab your back in return. As a result, the brain has become quite good at cleanly dividing *us* from *them* and feeling more empathy for *us*.

This may have worked great for our ancestors, but it's not so wonderful in today's world. Modern humans are diverse and culturally integrated; we're constantly surrounded by people who are unlike us. In a world like this, our brain's outdated empathy software becomes quite problematic. When someone appears different, our brain doesn't want their emotions to bleed into ours. We are unfortunately wired to feel apathy for many of the people we find ourselves surrounded by today.

This depressing theme has been quietly lingering behind the curtain of this chapter since the beginning. Remember that study of the firewalkers, where the onlookers in the crowd showed synchronizing heart rates while their relative walked the hot coals? Well, I left something out: This synchrony was *not* seen in the onlookers who were not related to the firewalkers. These neutral spectators sat comfortably, watching with a steady heart rate. Due to their smaller self-other overlap with the firewalker, their emotions remained disengaged.

Your brain sees your family members and friends as your "tribe." They belong to your in-group, which is why I had you walk through the jungle with them earlier in this chapter. In contrast, someone can be categorized as an out-group member for a number of reasons. For instance, race. When Black or white people are shown videos of hands being jabbed with a needle, they show greater activity in the insular cortex when viewing the hand that matches their race. Remember, the less overlap there is between you and another person in that Venn diagram, the less empathy you're liable to experience for them.

Religious identity is another variable in the equation of self-other overlap that can influence empathy. In a similar study, neuroscientist Dr. David Eagleman had subjects view photos of a hand being poked with a needle, where the hand was accompanied by a text label indicating a religion (like Hindu, Christian, or Jewish). The activity in the anterior cingulate cortex and insular cortex—those brain regions that represent the emotional experience of pain—was highest in people viewing the hand with *their* religious identity. The brain's empathy systems are literally recruited more strongly for those who belong to our religious in-group, and less so for those of different faiths.

What else can you think of that could make someone *different?* Unfortunately, many things may come to mind. It's something that worries me a bit in our modern age. In recent years, it seems like we've become quite good at identifying out-group members. We notice when someone's identity doesn't jibe with ours: They vote differently, dress differently, worship differently, or even eat a different diet. For each of these categories (and many others), in-groups and out-groups can be established, and your brain will likely favor those who share your group identity. Indeed, people report more empathy for people who share their sexuality or political ideology, and even *meaningless* identities can influence our empathy systems. When Dr. Eagleman randomly assigned participants to be either "Justinian" or "Augustinian"—two completely arbitrary groups—their brains showed those same patterns of tribalism, engaging empathy circuits more strongly for those on their team.

In a world that is increasingly diverse and heterogeneous—where differences between people have become emotionally charged and polarized—I fear that we may be losing our empathy, or at least narrowing it. With more and more reasons to exclude others from our tribe, we may be becoming less supportive of one another. This is a dark road, and one that I believe ends with a very dysfunctional society.

For example, imagine you drive past someone standing beside a broken-down car in a remote area. On a scale of 1 to 10, how likely are you to pull over and offer help? Really, pick a number, and listen to your gut. Now imagine their skin is a different color from yours, they have a bumper sticker supporting a politician you hate, and they're wearing a chain with a religious emblem that you don't identify with. Once again, how likely are you to help on a scale of 1 to 10?

If you were completely honest with yourself, your second number was lower than your first. That doesn't make you a monster. It doesn't make you racist, xenophobic, or hateful; it just makes you a human with a typically functioning brain. Your second number being lower is merely a reflection of your brain's organic impulse to protect those closest to you, based on evolutionary roots that are now irrelevant. It's yet another example of the *natural social pitfalls* driving us apart. While this tendency to empathize less with out-groups may be automatic, I believe it's what you do *next* that makes you who you are. You can recognize this gut impulse, make an internal note of it, and still pull over and help. It's decisions like this that make our communities stronger and show what it truly means to be human.

I believe we can take hold of our empathy through deliberate action and thought. In fact, there may be a simple way to do so. The next time you see someone in a tough situation, spend a moment wrapping your head around their experience. Engage cognitive empathy to understand what they must be thinking, and imagine the emotions that would accompany that headspace. Once you have a firm cognitive grasp on those emotions, try to immerse yourself in them. How would you feel in that situation? Imagine yourself there, in their shoes.

If you're having trouble, think back to self-other overlap. How closely does your circle overlap with theirs? If there's very little overlap, look for things that might shove them closer together. Maybe they're a parent, just like you. Perhaps they watch the same TV shows you like or have the same kind of dog. If that's not doing the trick, imagine swapping them out for someone close to you, like a best friend, sibling, or child. Now can you embody those emotions?

It matters to me that we have an empathic world. Empathy is like a flame that burns hot in our seats, forcing us to get up and help others. Should we really treat people with less humanity just because of the evolutionarily predetermined tendencies of our brains? I think not.

In a post-interaction world, let's not allow the brain's ancient tribal instincts to divide us further. There's no question that the world gets better when we extend our hands to one another, but our modern cultures are changing at a pace much faster than evolution could possibly keep up with, leaving us with an outdated system in an increasingly blended world. Today I implore you to ask yourself what kind of human you want to be. One who allows the archaic customs of a long-gone world to misguide you into isolation? Or one who can outpace evolution and outsmart the neural hardware you've been given, even if it means recognizing your brain's internal shortcomings? If you hope to improve your well-being through social connection, the choice should be clear.

## Key Takeaways

1. Emotions can be "contagious," passing between people through observation and mimicry of social cues.

2. Cognitive empathy involves *understanding* someone else's emotions and internal state, while emotional empathy involves *sharing* their emotional state.

3. Empathy is a tremendous strength. It was favored by evolution because it allows us to immediately grasp what others are thinking and feeling.

4. Empathy is largely shaped through early life experience, but it can also be modified by genetics. With that said, it's not hard-set in adulthood—it remains flexible to experience.

5. Empathy often involves modeling someone else's brain state, such as in empathy for pain.

6. The amount of empathy you feel is sensitive to many social and environmental factors. People tend to feel most empathy for those who are more like them.

To view the references cited in this chapter, please visit benrein .com/book.

# 5

———— ∘ ————

# ANYTHING YOU CAN DO, I CAN DO BETTER

Animal Interactions and What
We Can Learn from Them

In the midst of her PhD research, neuroscientist Dr. Monique Smith made a stunning discovery that completely changed the course of her career. She was studying alcohol withdrawal in mice when one of her experiments produced an unexpected result, revealing something astonishing.

Alcohol withdrawal is what happens when someone stops using alcohol after a period of intense consumption. It is far from pleasant; in fact, it can be legitimately painful due to something called *hyperalgesia*, a phenomenon where one's sensitivity to pain increases, sometimes dramatically. As a result, physical sensations that aren't typically painful can become uncomfortable and cause a great deal of distress. Dr. Smith was trying to figure out what causes this on a biological level by studying the brains of mice going through alcohol withdrawal. But to do this, she first needed to get them hooked on alcohol.

Luckily, this was no challenge. When she placed a bottle

containing 3 to 10 percent alcohol in their home cage (similar to the potency of a beer), the mice went on quite a bender. On a typical day, they each took in the mouse equivalent of forty-six drinks for a human weighing 150 pounds. Clearly they enjoyed it. Then Dr. Smith would take away their beloved juice to induce withdrawal and thereby lead them into that prickly state of hypersensitivity.

An easy way to measure this sensitivity in mice is called the Von Frey mechanical sensitivity test. Picture this: A mouse is standing on a wire mesh grid so that you can see and touch their paws from below. You pick up a small tool that looks like a toothbrush but with only a single thin bristle—not ideal for maintaining hygiene, but perfect for this experiment. Gently you poke the bristle into the mouse's back left paw. If the mouse flinches, lifts the paw, shakes it, or licks it, that's its way of saying, "Hey, I felt that, and I didn't particularly like it." If the mouse doesn't react, you move up to a thicker, heavier bristle. Eventually the mouse responds, allowing you to pinpoint how sensitive their paw is. Dr. Smith could tell that her mice were suffering from hyperalgesia because they reacted to the smallest, lightest bristles. However, there was a big problem: The mice in her *control group*—which hadn't touched a drop of alcohol—were *also* hypersensitive. What the hell was going on here? How did they, too, have hyperalgesia?

"At first I thought I'd made a mistake," Dr. Smith told me. "Maybe I gave them alcohol by accident." But rather than dismissing this confusing finding as a simple screw-up or a technical issue, she focused on it. She began to wonder whether the result was real. What if the control mice were hypersensitive because they were experiencing a form of emotional contagion? Could it be that they were taking on the hypersensitivity of their fellow mice, just as I absorbed the patient's panic during my fake dental exam?

It was an out-of-the-box hypothesis. Sure, it seemed a bit preposterous, but crazier things have happened. The control mice *were* living in the same room as those going through alcohol withdrawal, in open cages where smells and sounds could carry freely.

Maybe the control mice were sensing some signs of pain from the animals nearby and taking on their discomfort.

She had to test it. Dr. Smith ran the experiment again, but this time she kept her control mice in a separate room to prevent them from sensing any smells, sounds, or social cues that might be emanating from the mice in withdrawal. Remarkably, this solved her problem: Her control mice no longer showed the same hypersensitivity. The outlandish hypothesis seemed to be proving correct. The mice were somehow taking on the uncomfortable experience of their nearby friends. She had resolved the issue with her experiment, but how could she simply go back to studying alcohol withdrawal? What she'd found was simply too intriguing to ignore.

If the mice were truly taking on one another's state, then this was something resembling empathy. It's a bit like the example from the last chapter, where your partner stepped on a thorn and you took on a corresponding sense of discomfort. Is this what was happening in the mice? Were they taking on a bit of one another's pain?

To determine if so, she ran the experiment again, but this time with mice experiencing true *pain* rather than alcohol withdrawal: a bad case of arthritis and inflammation in one of their paws. When she housed these arthritic mice in the same room as healthy controls, the effect was clear. Within a few days, the healthy mice began acting as if they, too, were suffering from arthritis. They flicked or licked their paws at the gentle touch of a thin bristle. Something remarkable was happening here.

Dr. Smith's discoveries teach us something profound: that mice can sense and share one another's emotions. Somehow these small, simple creatures—which most humans view as appalling vermin, something very distant from us—can experience empathy just like we do.

She named the phenomenon the *social transfer of pain* and pivoted her research to focus on studying it. After finishing her PhD, she moved on to Stanford University to explore what was happening in the brains of these empathizing mice. What she found confirmed her suspicions: The social transfer of pain required brain areas that are linked to empathy in humans. When healthy mice interacted with arthritic mice, they showed more than *twice* as much activity in their anterior cingulate cortex than when interacting with a healthy mouse (yes, that's the same brain area found in human studies of empathy). When she silenced the anterior cingulate cortex using a tool called *optogenetics*,[+] the mice stopped taking on one another's pain, suggesting that this region was necessary[+] for their empathy-like behaviors. However, it turned out the anterior cingulate cortex wasn't acting alone. It was sending crucial signals to the nucleus accumbens—that M&M-like brain area involved in motivation and reward. Dr. Smith had seemingly discovered a significant new relationship between these two brain systems, suggesting that they work together to drive empathy. Moreover, evidence was building that she had truly uncovered a form of empathy in mice, challenging the notion that humans are the only ones who wield this valuable power.

I remember reading Dr. Smith's research article in amazement, sitting at my desk nearly three thousand miles away from Stanford in Buffalo, New York. It was one of the coolest and most unbelievable studies I had ever seen. Little did I know that I would soon ship off to work in Dr. Rob Malenka's lab at Stanford—the same place Dr. Smith had led these studies. It was there that she and I would cross paths and unexpectedly team up to trailblaze another study on empathy . . . However, that's a story for chapter 9.

⊚

You could certainly argue that human empathy is deeper and more complex than what Dr. Smith found in her mice. I mean, we do more than just see the world from others' perspectives and appreciate what it might feel like; we also step up and *do something* about it. We are often motivated to offer help or console our struggling peers. This is a step above empathy, no?

As a matter of fact, this is what researchers call *compassion*.

Compassion goes a step beyond empathy in that it incorporates *a desire to help*. If we think of caring for others like a staircase, compassion would be the top step.

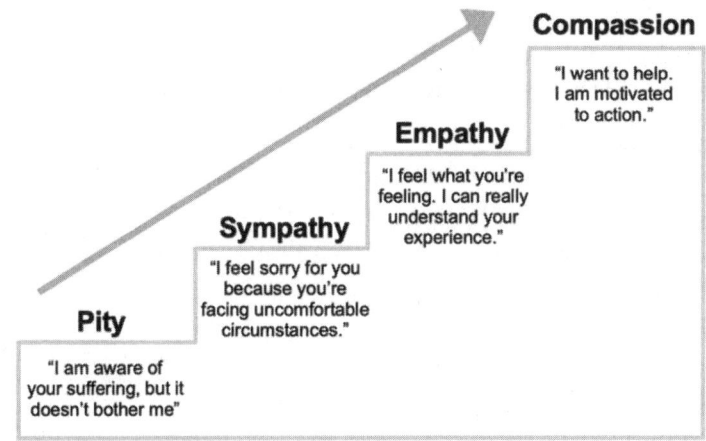

**Compassion**
"I want to help. I am motivated to action."

**Empathy**
"I feel what you're feeling. I can really understand your experience."

**Sympathy**
"I feel sorry for you because you're facing uncomfortable circumstances."

**Pity**
"I am aware of your suffering, but it doesn't bother me"

At the very lowest step of such a staircase would be *pity*: an awareness of another's misfortune, but with an emotional disconnect. Pity is when you recognize that someone or something is suffering, but you don't feel bad. When a wasp is trapped in your house buzzing interminably against a window, you know that it's facing an unpleasant situation, but you don't really feel bad for it. You're not emotionally attached to the outcome. You pity the thing.

One step up would be *sympathy*. Sympathy is when you understand someone's circumstances and you *do* feel bad for them. This sorry feeling is what distinguishes sympathy from pity. However, sympathy doesn't involve *sharing* their exact emotions like in empathy.

*Empathy*, as we've discussed extensively, is the sharing of another's experience. It's when we understand and take on others' emotions.

Finally, *compassion* is the top step, and the only step that involves a desire to act in response. It's characterized by a motivation to alleviate someone's suffering. We humans take great pride in this. We rescue sick animals and nurture them to health, offer hot meals to those in need, and enter burning houses to save our neighbors. When we exercise compassion, we put our humanity proudly on display. *This*, you may argue, is what the mice lack.

But is compassion truly a *human* quality, or can animals possess it, too? Is it possible that Dr. Smith's mice could have gone a step above empathy, or had they reached the upper limit of their social capacities?

Although we humans may like to think of ourselves as the sole proprietors of compassion, we are but one of the many species on Earth who wield this power. There are many documented cases of animals showing compassion, with numerous published scientific papers reporting compassionate behaviors in the likes of zebra fish, primates, elephants, and so on.

Yes, you read that right—I used *compassion* and *zebra fish* in the same sentence. Zebra fish are tiny little critters that hail from the minnow family, and indeed, they show behaviors that resemble compassion. When a zebra fish gets scared, the others nearby tend to hang out more near them. This is thought to be the zebra fish version of *consolation*—something we humans do all the time. For instance, when someone breaks down in tears before you, you might console them by hugging or holding them. Offering this physical touch can help soothe their emotional pain, and swimming nearby might be the equivalent for a zebra fish. It makes sense when you think about it, because fish like to swim in schools for the sake of their safety. Clustering around a distressed fish may ease its discomfort by making it feel safe and protected.

Now, I wouldn't blame you if you're unconvinced—but here's the kicker. When researchers genetically engineered the zebra fish to lack the genes for oxytocin or the oxytocin receptor, they stopped hanging out near their petrified pals. It seems that even organisms as basic as a *minnow* are capable of compassion, and they're driven by the same molecule that the human brain uses for connection.

If zebra fish can do this, then it should come as no surprise that rodents, too, have been shown to console one another. Prairie voles are a unique type of rodent in that they form monogamous pair bonds. When a male and female prairie vole tie the proverbial knot, they're partners for life. It's even been reported that when the female partner dies, the male will not look for a new partner. Who's cutting onions?

As you might expect from such a wholesome creature, prairie voles have been shown to console their partner after stressful experiences. After a vole goes through a stressful experience, their partner will gently lick them to ease their distress. Voles will do the same for a sibling—but interestingly, not for a stranger. It seems that just like humans, they may reserve their empathy and compassion for those closest to them. What's more, blocking oxytocin signaling prevents the voles from consoling one another, and blocking it *only* in the anterior cingulate cortex (that same empathy-related region) is enough to restrain their compassion.

Perhaps we wouldn't have been wrong to expect compassion from Dr. Smith's mice. Indeed, mice and rats have been caught doing incredibly kind and altruistic things. For instance, when rats are uncomfortably hoisted in the air by a harness, their buddies will press a lever to lower them to safety. Another study gave rats a difficult choice—open container 1 to liberate a restrained rat or open container 2 to access some chocolate. What do you think they did? What would *you* do?

The rats did the nicest thing possible. They opened both containers and then shared the chocolate. Are you kidding me? Where did these rats learn their manners?

Let's be honest with ourselves: These behaviors are a glowing example of what humans should strive for. If your kindergartner chose to free their classmate and shared the chocolate, you would be a *very* proud parent. But these aren't kindergartners we're talking about; they're rats.

It's one thing to detect another animal's distress and offer help; it's another to share chocolate for no reason. This seems to be going above compassion toward sheer *generosity*, and believe it or not, other research also shows that mice are capable of generosity. In one study, mice were given two options: (a) take a treat for themselves, or (b) get a treat *and* give one to another mouse. Shockingly, most mice preferred to make the generous decision. Even when the researchers made it *more* expensive to be generous—requiring more work to share a treat than to get one themselves—most mice still chose generosity.

I believe these studies reveal the true nature of Earth's creatures: they empathize and care for one another. This is somewhat of a new discovery in science; as recent as the 1900s, it was generally believed that most animals couldn't even feel pain. Now we see that they don't just feel pain for themselves but can *detect it in one another.* It certainly makes you wonder what other unchallenged notions exist today that we will mock in a century's time.

I also can't help but wonder, if emancipating one another from traps and sharing treats is standard behavior for many of Earth's beings, why we humans have seemingly lost so much of this. While these animals may fall short of us in their intellectual abilities, do they exceed our kindness and companionship? Have our higher-level abilities gotten in our way? In our journey to continue building, growing, and reaching new heights as a species, are we growing in the wrong direction, losing some of the fundamental social attributes that characterize even the most basic forms of life? Considering that **division is the enemy of brain health**, what the hell are we doing? *Why* are we so insistent on being divided? What do we get from the division we've embraced, besides a poorer mood, compromised health, and a less nurturing world?

⊗

In Nanterre, France—a few miles northwest of Paris—a group of scientists put six bottlenose dolphins through a generosity test very much like what was done in mice. The dolphins were given three choices: (a) be selfish (take a reward for themselves), (b) be generous (give another dolphin a reward while also earning one themselves), or (c) choose violence (*nobody* gets a reward!). Just like mice, the dolphins tended to prefer the generous choice. What's more, the researchers noticed something else intriguing. The dolphins were far more generous when sharing with a dolphin *of the opposite sex*. When given the chance to flirt, the dolphins chose option b nearly 70 percent of the time. Meanwhile, they shared their reward only about a quarter of the time with a same-sex recipient. Apparently, chivalry is not dead in dolphin culture. This models some human trends. For instance, men are willing to donate more to charity in the presence of a woman. Just like us, dolphins seem to think they can win over potential mates with kindness.

Perhaps even more interesting, the dolphins' kindness jumped dramatically when their offspring were present. With their young nearby, the dolphins chose to share the treat over *90 percent* of the time! In contrast, their generosity plummeted to about 20 percent when there was another adult dolphin watching. In animal research, it's always best to exercise caution and avoid *anthropomorphizing* (attributing human characteristics to animals). However, it's hard not to in this case. Why would a dolphin become more generous to other dolphins in the presence of its young? To me, there's only one answer, and it's undeniably human: to model and teach good behavior.

This again makes me pause and reflect on humanity. How often do we model good behavior for our young? When we're in line to cash out with our kids, do we let the shopper with just a couple of items cut the line ahead of us? Do we offer our bus seat to the elderly man to teach our kids about compassion? Dolphins presumably have less complicated social dynamics than human culture, but for some reason they put their best flipper forward when their children are watching. What do they have to gain that we don't? Have we slipped away from our biological instincts, falling prey to selfish impulses imposed by the pressure of modern life?

Elephants, too, show striking signs of empathy. They're one of few species (including humans) whose brains have those von

Economo neurons that are linked to empathy. In Southern Kenya, a population of over 2,200 elephants was studied for thirty-five years, during which many empathic events were documented. Over 125 reports detail elephants offering one another comfort, such as mothers comforting their young through physical touch. The elephants were often seen protecting those who were too young or too injured to defend themselves from predators. When a calf fell over, the adults would help it to its feet with its trunk. On fifteen separate occasions, a calf fell into a river or ditch. In response, an adult used its tusks to dig at the bank and create a channel for the calf to climb out on their own.

Each of these behaviors necessitates empathy. For a mother to comfort her calf, she must access his emotional state and recognize that he *needs* comforting. For one animal to protect another from danger, it must understand that they are too weak to protect themselves. For an adult to dig a channel that a calf could navigate, it must take the smaller animal's perspective and see the world through their eyes. This requires *theory of mind*: the ability to understand that others have their own thoughts that are different from your own. Remember when I confused Sandi at the garden store because she didn't see me read her name tag? In that interaction, theory of mind allowed me to see the situation from her perspective. In the case of the elephants, the adult had to imagine the world from the calf's perspective to figure out how the landscape needed to change to permit the calf's escape. Although it may seem simple, this is a quite advanced ability. In fact, humans don't develop theory of mind until about preschool age, and continue developing up through ages six through eight. It's remarkable to think that an elephant may be capable of something a human child has not yet mastered.

◯◯

At last we reach the top rung of the animal kingdom: primates. Monkeys and apes are our most closely related species, so there's little surprise in the fact that they display humanlike social behaviors. Chimpanzees kiss and embrace one another to reconcile after they've had an argument, and rhesus monkeys will refrain from awarding themselves a treat if it means another monkey will receive an electric shock. Monkeys also demonstrate theory of mind, as discovered by a clever study where researchers set up two chimpanzees in side-by-side rooms, divided by a clear glass barrier with a hole they could reach through. One monkey was offered some juice, but rudely, they weren't given a straw. On the other side of the glass divider, the other chimp was given a box of seven tools, which included, of course, a straw. Incredibly, the helper monkeys were able to pick out the straw and hand it over so their buddy could slurp up and enjoy the juice. This is a clear demonstration of theory of mind: The monkeys could figure out their partner's challenge and determine the proper tool for the job. If you have kids, set up this experiment in your home and see how they fare. If they're under four or five years old, they might struggle.

Another report tells an unforgettable story of compassion in monkeys. On a primate nature reserve in Southeast Brazil, researchers have been observing a protected population of black-fronted titi monkeys since 2003. For years, the scientists have documented that the monkeys exist in multiple factions that do not get along. In the words of the researchers, "only low levels of inter-group tolerance have been observed"—sort of like fans of rival sports teams.

In 2011, the groups began competing for resources, causing tensions between two factions to escalate. Fights would break out, and one male monkey was particularly noted for harassing a female from the other group, who attacked him in return. One morning not long after a fight, the researchers noticed the male was acting strange and seemed to be hurt. He was walking oddly, spending a lot of time lying down, and having trouble climbing trees. In his malaise, the injured monkey wandered near the rival group. Normally this would be met with confrontation—lots of yelling and combative behavior—but that day, the rival group let him be. The hurt monkey continued to follow the other faction wherever they traveled, and surprisingly, they allowed him. At several points when he paused to lie on the forest floor—clearly injured—the rival group even stopped to wait for him.

When nightfall came, the faction settled down to rest. Then the limping male boldly entered their camp and something extraordinary happened. He approached the female who had injured him, leaned against her, and intertwined his tail with hers. She began to groom him, even as her male partner watched without complaint. Then the pair separated, and the group drifted off to sleep. After this, the injured monkey was never seen again. Presumably, he had died overnight.

This story is both touching and tragic. In a time of need, the rival faction parted from their tribal instincts and offered aid and comfort to an enemy. By way of their empathy and compassion, he spent his final hours in the company of a group. They respected his boundaries and welcomed him into their camp, perhaps for the first time in all his life. I find it impossible to read this story and *not* believe that animals experience empathy and compassion. Like us, they are complex beings. They can understand others' emotions and adapt their beliefs in changing circumstances. Compassion is clearly not something we humans can claim as our own; rather, it's something that likely predates the emergence of our species. In other words, compassion evolved in other animals far before we came along. We merely inherited it.

Before we get carried away assuming that every animal alive is capable of empathy, here's a word of caution. Empathy and compassion may not be universal in the animal kingdom; there are probably some exceptions.

For example, chickens. One study tested for empathy in chickens by having them watch one another experience something unpleasant—a blast of air sprayed in their direction. When a chicken gets blasted like this, they show clear signs of distress like an increased heart rate. However, when one chicken watched another chicken get the puff, they would sit by unbothered, twiddling their claws. Apparently, adult chickens don't show any visible signs of empathy toward other adults. However, they *do* show empathy for their chicks. When adult hens watched their little baby chick receive a puff of air (which was harmless, by the way), they showed signs of empathy, like an increased heart rate and a decrease in eye temperature—a sign of stress in chickens.

It's easy to understand why a chicken would show greater empathy for their offspring than for a fellow adult chicken; after all, caring for one's young is paramount. We humans are also exceptionally sensitive to the state of our children. But unlike humans, chickens seemingly don't empathize with other adults.

I share this to point out that empathy likely differs between various species. After all, it's an evolutionary trait that evolved partly to promote caretaking behaviors. Different animals experience different evolutionary pressures, which may cause empathy to look different across species.

We humans may be among the *most* empathic species. This is a tremendous gift, and one we should honor. Empathy is a strength; in fact, I might even call it a superpower. I hope that we as a species will choose to sustain the evolutionary tradition of empathy and savor this gift from our past—lest we all turn to chickens.

⊙⊙

While it may seem like the purpose of this chapter was to disparage humankind or make you feel bad, that was not my goal. I wrote this chapter to point out something I find deeply meaningful: When biological forces coil together to create *life* on planet Earth—whether that's a zebra fish, a rat, or a titi monkey—that life tends to manifest in a way that is kind, compassionate, and even generous. I find it astounding that these many species, with dramatically different levels of intelligence, are all capable of empathy and compassion. You could argue that this drive for togetherness is present even at the most *basic* levels of life: individual cells. When scientists put heart cells in a dish together, they self-organize into clusters and begin *beating* in unison. Brain cells in the right conditions will assemble into groups, forming tiny brain-like structures. If the default state of biology is *unity*, why have we become so *divided*? How often do we fail to show our own kind the respect that zebra fish show fellow zebra fish? How much does our intellect get in the way of this basic biology, holding us back from reaching those higher steps of empathy and compassion?

I think this is a good reminder to ask yourself where you stand. Would you go out of your way to save a stranger from harm or to console a hurt member of an opposing political party? Would you share money or food with someone else when it's less work to just keep it yourself? We began this chapter by asking whether animals are capable of human behaviors like empathy and compassion, but perhaps we should now ask ourselves where *our* humanity stands. Do we, as a civilization, still embody the altruistic, prosocial tendencies that we so proudly consider *human*? Do we really deserve to claim ownership of those traits?

To be the best, healthiest version of humanity, we must acknowledge the barriers that divide us and find ways to overcome them. We are blessed with tremendously powerful brains, and we should take pride in this. *We* are humankind. We are the *pioneers*, the *builders* of the Earth, but we seem to be letting this advantage get in our way. Our powerful brains have been plopped into a modern culture that is far different from our ancient origins, and they're struggling to navigate it, preventing us from achieving what biology originally intended for us. In our endless quest for innovation, our steadfast determination to grow in every direction possible, we seem to be growing apart. We deserve better, and we can do better.

What truly distinguishes us from these animals is our intellect. How amazing is it that we have the ability to *study ourselves*?

We can quantify what's happening in our own brains and communities and use this information to build a better world. *This* is how we should be putting our advanced intellect to work. *This* is something that minnows, rodents, dolphins, elephants, and monkeys can't claim. *This* is what makes us human—overcoming obstacles and advancing in the face of adversity. We should honor the biology we've been gifted by constructing systems that bring us back together and let us savor the health benefits of social connection.

*This* is how we put the *human* back in *humanity.*

## Key Takeaways

1. Mice have been shown to take on one another's pain, a process that resembles empathy and involves empathy-related brain areas.

2. Pity, sympathy, empathy, and compassion are distinct. Pity involves the least emotional concern, while compassion involves the most.

3. Animals ranging from zebra fish to elephants show signs of empathy and compassion.

To view the references cited in this chapter, please visit benrein .com/book.

# 6

## A VIRTUAL WORLD

### What Do We Get from Online Interactions?

Around 500 BC, Queen Atossa of the Persian empire wrote a consequential historic letter. It wasn't the letter's content that made it so famous and impactful; in fact, historians don't even know what she wrote. Atossa's letter is noteworthy for a different reason: It's thought to be the first letter ever written.

When the queen put pen to paper, it changed humanity. People wanted to be like her, so they started writing letters, too. It "established the genre" of letter writing, according to history professor Dr. Bríd McGrath. What the queen and her disciples didn't realize is that this was the first time in human record that a form of communication not involving face-to-face contact had emerged and become mainstream. Surely Atossa had no idea what she'd done. She'd triggered an inevitable shift in the way humans would continue to interact, hurtling us more and more toward nonlocal forms of interaction over the next several thousand years.

Roughly 2,400 years later—in the 1880s—the telephone was developed. This time, humanity unlocked a new way to interact remotely through speech instead of written language. Walkie-talkies were invented shortly after in 1937. The first email was sent in 1971. Then in the early 2000s, we saw the rise of social media platforms like Myspace, Facebook, and Twitter, followed later by Instagram, TikTok, and so on. Nowadays, video-based communication platforms are the norm, with Zooms and FaceTimes built into everyday activities. At last, we've reached the inevitable destination of the cascade Queen Atossa set off. Or perhaps more changes await us in the years to come.

In retrospect, these massive strides were inevitable for humankind. Why *wouldn't* we progress from handwritten letters on papyrus to 4K video streams on ultrawide monitors? It's just what we do. But was the human brain ready for all this? Is it ready today?

Recall that our brains were shaped to be social because living in groups kept us alive. Fossil records show that our ancient humanlike ancestors were around at least 2 *million* years ago, but until Queen Atossa's letter, almost all human interaction was carried out face-to-face. That means that for the first 99.9 percent of humanity's existence, the hands of evolution shaped and optimized our brains for in-person contact. We grew exceptional at understanding one another *in real life*—not over video calls or social media feeds.

The evidence for this is written all over our bodies. Evolution equipped us with all sorts of handy tools and systems for face-to-face interaction. If you don't believe me, just look in the mirror. See how most of your eyeball is white? That white part is called the *sclera*, and the fact that it's white makes us unique. In most animals, sclerae are dark. For humans, they come standard as white. What for?

This is the work of evolution. Lining our eyes with bright white indicators makes it easier to tell where someone is looking, and this helps us read one another's minds. Just think, if a sibling approached you with greasy fingertips from scarfing down potato chips and looked at your T-shirt, you might say, "Don't even think about wiping those greasy fingers on me." You could tell what they were thinking by simply looking at their eyes. A recent study found that both humans and chimpanzees were better at reading where human eyes (with white sclerae) were looking versus chimpanzee eyes (with dark ones). When the researchers edited photos of chimpanzee eyes to make their sclerae white, both humans and chimpanzees could more easily tell where they were looking.

This ability to read eyes makes us more effective when working in teams. For example, hunting in groups is much easier if we can intuit our partner's next move by simply looking at them. Interestingly, certain dogs have apparently evolved light-colored irises for the same reasons. Scientists have found that canines who live and hunt in groups are more likely to have brighter and more visible eyes, making it easier for them to work efficiently in teams. Bit of a frightening thought, isn't it? Being hunted by a pack of deadly animals that can communicate by simply looking at one another . . .

Evidently, human beings have been shaped for social contact both inside and out. Our brains reward us for connection (and punish us for isolation), and our bodies have tools to help us communicate. However, these tools are meant for *face-to-face* contact. As a neuroscientist, I think about this observation frequently, but it seems to be missing from the conversation about loneliness and connection. We blame technology, social media, and Zoom calls for our loneliness, but we may not be appreciating exactly *why* they are the problem. The science of the brain gives us unique glasses through which to view this. What happens when we interact online, where our white sclerae aren't visible? What becomes of these prehistoric tools in a modern virtual world? And is technological interaction better than no interaction at all?

If you're still in front of a mirror, look above your eyes to those handsome strips of fur we call eyebrows. Modern humans have shed much of the fur our ancient ancestors had, but for some reason we still have eyebrows. That might be because eyebrows shield the eyes from damage in some circumstances (like catching rain or blocking sun), but they also offer a hidden advantage in communication. Eyebrows are important for expressing emotions, and the fact that we have such robust muscular control over them says a lot. A raise of both brows can signal surprise or joy, while lowering them may indicate anger or sadness. Even in the simple black-and-white faces below, look at how the inclusion of eyebrows dramatically changes the expression. Can you feel the emotional differences?

Our expressive eyebrows and white sclerae are visible hints of our history. They tell a story of a past that's rich in social contact, where our brains and bodies evolved to maximize face-to-face information transfer. But after millions of years, that's changing.

Much of the time we used to spend face-to-face is now spent face-to-screen. In 2022, the average American spent four and a half hours per day using their phone, and much of that time included virtual interactions. But you can't hear your friend's tone of voice when you text him. You can't see other users' eyebrows on Twitter. You can't smell Grandma's perfume when you FaceTime her. Many of the social cues that enrich our interactions and help us understand others are totally left out in virtual interactions, and I believe that's a problem.

Unlike our phones and computers, the brain doesn't get software updates. It's still the same old piece of machinery that Queen Atossa had: an ancient device that originally took shape in much different circumstances. The brain on social media is like a horse-drawn carriage on the autobahn. It's out of place, an archaic tool tumbling its way through a brand-new, fast-moving, and often overwhelming virtual world. So what is the result?

## THE EFFECTS OF ONLINE INTERACTIONS

In chapter 1, we established that interacting with others activates reward signals in the brain and brings us joy, even if it's just interacting with a stranger on a train. In chapter 2, we learned that isolation is destructive to our well-being, and that division is the enemy of brain health. There's no doubt that we live in a divided world, where so-called interactions can happen on a phone or computer screen between two people thousands of miles apart. These virtual social experiences are almost like a fusion of interaction and isolation. Yes, we're interacting . . . but we're not together. What does the brain make of these moments? Do they offer valuable connection that can enrich our brains, or is chatting through a phone more like being isolated?

Let's first acknowledge that there are many different ways to interact virtually, and they each exclude certain social cues. When we communicate in digital formats, some of the signals that typically enrich our face-to-face conversations are filtered out—like body language, vocal tone, facial expressions, pheromones, and other cues. As a result, the texture of our interactions is flattened.

The closest thing to a real interaction is a video call, on Zoom or FaceTime, for example. (Note that virtual reality interactions are more lifelike, but I'm skipping them because they're not commonly used in today's world . . . *yet*.) In the move from real-life interactions to video calls, we lose several valuable cues. First, eye contact becomes impossible. Second, we can't smell *olfactory cues* (odors) that might be coming from our social partner. Third, we may potentially miss out on body language, depending on how much of their body is visible on-screen. If our partner has low-quality video, audio, or internet connection, we may also lose out on some facial expressions and vocal tone.

The next step away from real-life encounters would be phone calls, which *definitely* filter out facial expressions and body language, in addition to eye contact and olfactory cues. When we speak on the phone, all we really have to work from is vocal tone.

Finally, *text-based interactions* are the next step away from real-life interactions. Texts and emails lack vocal tone, leaving only our naked words to convey thoughts and emotions. I would thus rank interactions in the following order: **in-person > video > phone > texting/messaging.**

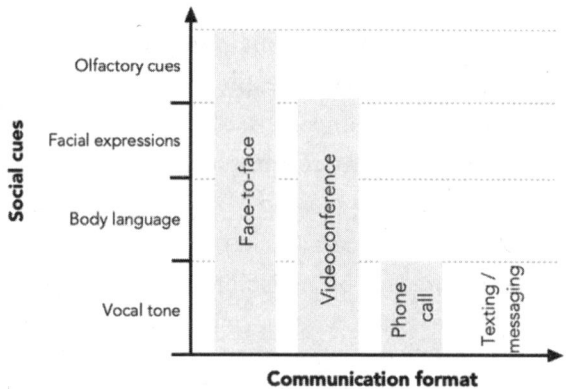

As we move from left to right in the above graph, we progressively strip away the substance and texture of our interactions, leaving behind an incomplete representation of real-life exchanges. Are the remnants even meaningful? Do they offer any of the benefits of in-person interactions?

Unfortunately, the outlook is a bit gloomy.

After experiencing less "lifelike" interactions (like texts or phone calls), people tend to feel lonelier, sadder, less affectionate, less supported, and less happy than after interacting in person. People report the most enjoyment from talking in person, and the least from interacting through text. So as we move from left to right on this graph, our interactions do seem to become less rewarding and pleasant. One study found that in-person interactions were 4.5 times more effective at boosting well-being than virtual interactions. Does this mean in-person contact more potently activates the brain's social reward systems like oxytocin, dopamine, and serotonin? That exact study hasn't been done yet, but it wouldn't surprise me if it were the case.

Luckily, there is a silver lining: Interacting online was still better than having no interaction at all. To me, this seems like good news and bad news. The bad news is that virtual interactions fall short of in-person gatherings, failing to offer the same boosts in mood and sense of social support. But the good news is, online interactions are not entirely devoid of value: They still leave us feeling better than talking to nobody at all. It seems that by *stripping away* the meat of our interactions, we are eliminating much of the rewarding value and potentially ushering in other issues.

Comparing the top step of our graph to our bottom step, we see major differences. Obviously, texting is very different than meeting face-to-face. However, the drop in quality may not be so precipitous as we go only one or two steps down. Studies show that video and audio calls don't differ so much from in-person conversations in terms of how people enjoy them or how close they feel with their partner. When people meet over video, they also show *facial and linguistic mimicry* (mimicking facial expressions and language style, respectively). This is promising, as it suggests that phone and video calls may still convince our brains and bodies to engage with one another in certain ways we would in person.

Interestingly, other evidence suggests that even text-based communication can be powerful in the right conditions. A 2024 study found that mothers and daughters can achieve *interbrain synchrony* while texting. Interbrain synchrony is a remarkable phenomenon that we'll talk more about in chapter 7, but to put it briefly, it's when two people are interacting and their brain activity synchronizes. It happens more readily in people with strong relationships, like parents and children. The fact that mother-daughter duos showed interbrain synchrony over computer messaging is encouraging, suggesting that text-based interactions can still be meaningful. However, the pairs showed greater interbrain synchrony when interacting in person, again indicating that texting falls short of the real deal.

There are several key learnings we can take away from this research. First, I believe that we can't equate virtual interactions to in-person ones. The science doesn't support it, and at a basic level, our direct experiences don't support it. In a divided, post-interaction world, I think we should be prioritizing in-person connection to maximize our well-being. Chatting over text, email, FaceTime, or phone leaves out important social cues that enrich our connections; these virtual forms of interaction are incomplete and may not thrill the brain quite the same way. However, if you're forced to interact nonlocally, a phone or video call is likely better than texting.

In my personal life, I try to prioritize in-person interactions whenever I can. I jump at opportunities to hold work meetings in person instead of on Zoom. The other day when a friend asked if I had time for a phone call, I invited him to my house for coffee instead. Our bodies are built for in-person connection; we have all sorts of incredible tools and abilities embedded within us. So why not use them?

⚭

In a world where just about everyone has a computer in their pocket at all times, a new form of interaction has emerged: a sort of *in-between* state where people are interacting in person but also are using their phone.

We've all sadly experienced this situation. You're in a conversation over dinner, and suddenly you find yourself talking to the top of a head. Your guest has turned their attention to their phone, and all you're getting is an occasional nod and grunt. Boy, that sucks. There's a name for this: *phubbing*. The word, coined in 2012 by an Australian advertising agency as part of a marketing campaign, combines *phone* and *snubbing*, and according to several studies, phubbing sucks the value out of social interactions.

In one of these studies, three hundred people sat down in a café with friends or family. Half of them were instructed to put their phone away, while the other half were allowed to simply leave it on the table. Those who left their phone out reported *less enjoyment* from the interaction and *felt more distracted*—just from leaving it on the table! Another study confirmed this, revealing that millennials felt worse and less connected after they'd used their phone during a face-to-face interaction. Moral of the story: Put your phone away. It's an easy way to make the most of your interactions and avoid distractions.

## IS SOCIAL MEDIA EVEN *SOCIAL*?

The last few decades have been characterized by a massive surge in social media use, and there's growing concern that this may be causing harm. But should we really be worried? Well, when Facebook was rolled out across American colleges, there was a corresponding decline in the mental health of students at those schools. After over four hundred schools in Norway banned smartphones, psychological symptoms in girls dropped by 29 percent, and bullying in both boys and girls fell by 43 percent. When people take just a one-week break from social media, they report greater well-being, less anxiety, and lower depression scores. Even limiting social media usage to thirty minutes a day for two weeks has been shown to improve well-being and reduce anxiety and depression.

I find this a bit perplexing. After all, the whole foundation of this book is that socializing is terrific for health. However, "social" media seems to have a net *negative* effect on well-being. Moreover, when the people in these studies took a break from social media, they reported feeling *less lonely*. Quite ironic, isn't it? Spending time on so-called social media leads to feelings of *loneliness*. It's almost like the entire premise of socializing is reversed.

This begs an obvious question: **Is social media even *social*?**

It certainly doesn't seem to be. If social media were equivalent to real interaction, you would feel better after swiping on these apps. Your mood would improve, and you would feel less lonely. Closing down the apps would feel like being *isolated*, accompanied by a rise in cortisol and a decline in mood. However, the exact opposite is true. Using social media makes us feel lonely and sad, while abstaining from it leads to positive effects.

In truth, there are many factors at play here. Studies point to issues like body dissatisfaction and social comparisons as potential drivers of social media's effects. As they say, "Comparison is the thief of joy." Everyone knows that social media is not an accurate representation of real life. On it you see people at their best— sometimes even edited to look better—often sharing only what they're most proud of. Social media can be treacherous, especially for young people forming their identity and working to fit into social structures. This certainly contributes to its effects on mood.

We must also recognize the role of sleep. Teens who interact more in person have better sleep quality, while those who interact more *online* get worse sleep. It has been shown that this sleep deficit is one of the key factors driving social media's negative influence on well-being. Why? Well, people who use social media more tend to do so at night, staying up late scrolling. As a result, they get less sleep. Sufficient sleep is a core pillar of mental health. In fact, if you get less sleep, it's almost guaranteed that you will be less happy. When you spend your nights on social media, you're certainly not doing yourself any favors.

Nighttime social media use can also hurt you for another reason: screen exposure. Your eyes are a part of your nervous system, and they feed in directly to your brain through the optic nerve. This nerve carries information about light in your environment, which the brain uses to regulate your circadian rhythm. When you hold a screen just inches from your face, the artificial light streaming through your eyes tells your brain that it's daytime. This can suppress your body's natural melatonin production and impair your sleep quality. Thus I strongly recommend limiting phone usage before bed, even ideally putting your phone away a few hours before bed.

An easy way to implement this healthy change is to leave your phone outside the bedroom at night; just plug it in to charge in another room. If you use your phone for alarms, just buy an alarm clock. Seriously, I found one on Amazon for $9.99 with good reviews. Trust me—the quality of your sleep is worth so much more than $9.99. High-quality sleep will change your life; but that's a topic for another book.

So to answer the question *Is social media social?* I would say probably not. There's no doubt that social media is not equivalent to being with others in real life. They have opposite effects on well-being, sleep, and even loneliness. I generally believe that social media is not very good for us, but of course it can be very pleasant and valuable in moderation. With regulated use social media can be a lovely way to pass time and gather information, but under no circumstances should it be used as a replacement for real social interaction.

## BAD FOR BUSINESS?

If we compare contemporary life with the world of ten years ago, one of the biggest differences we can see is the prevalence of remote work and virtual meetings.

As we emerged from COVID-19 lockdowns, these were some of

the adjustments that we kept. Why wouldn't we? Instead of having to get dressed, drive to work, deal with parking, and be stuck sitting through boring meetings, workers can roll out of bed, stay in their pajamas, and join calls while sipping a cup of homemade joe. So this form of work stuck. But is it any good for us?

The science on virtual meetings is new and emerging, as we haven't had many years to study them. However, the early evidence isn't looking favorable. It appears that they fall short of real-life meetings in a few ways.

First, people tend to feel more fatigued after virtual meetings. Perhaps it's more exhausting having to awkwardly take turns speaking, mute and unmute every few minutes, and keep an eye on your background to make sure nothing wild is happening behind you (like your dog throwing up or your kid walking around nude).

A second downfall is that virtual meetings may suppress creativity. The evidence for this comes from a study in which subjects who were paired up online to brainstorm creative uses for a product came up with fewer ideas than those who met in person. To see if this is also true in a real-world setting, nearly 1,500 real-life engineers were asked to do the same. Once again, the engineers who teamed up online came up with fewer ideas than those who met in person.

Why is this? Are we more easily distracted when we meet online, getting caught up in our surroundings? Actually, it turns out to be the opposite. When we video chat with others, we're required to collapse our attention into a small box on a computer screen. This *narrowing* of our visual focus may narrow our *cognitive* focus as well, limiting us from literally thinking outside the box. The researchers came up with a clever way to test for this: During the next round of meetings, they stashed unusual props around the room. It turned out that people who met online spent less time looking around the room and consequently remembered fewer of the props. In contrast, those who *did* spend more time looking around the room (and remembered more props) came up with more creative ideas. This suggests that when we concentrate on a box on a computer screen, the window of our imagination may tighten, too, thereby restraining our ability to think creatively. Perhaps when we brainstorm with groups online, we should encourage participants to get up and move around or otherwise attend to the room around them. By broadening the scope of our attention, we may be able to broaden the scope of our ideas and collaborate more effectively in virtual settings.

At least when everyone's paying attention . . .

Distractions can be abundant in virtual meetings, and sometimes the temptation is irresistible. If you aren't locked in a real-life conference room with real-life coworkers, it's easy to find yourself typing an email or doing some other work. A study of Microsoft employees during COVID-19 found that multitasking was rather common in virtual meetings: About 30 percent of the time employees were in a Zoom call, they were simultaneously sending emails. The employees were most likely to multitask when meetings were longer, comprised of larger groups, and held in the mornings. In contrast, they were less likely to multitask when meetings were held on Fridays.

This may be explained by one simple and obvious fact: People are more likely to multitask when they have a lot to do. So when do people feel this pressure? Probably when they arrive at work to a full email inbox, which may be why employees are more likely to multitask in morning meetings. Similarly, when a meeting goes too long, we may start to stress over other work and quietly turn to work on those projects. Meanwhile, the workload is typically lightest on Fridays, so employees are less prone to getting sidetracked during meetings held then. Maximizing focus in virtual meetings may be a matter of balancing workload and keeping other pressures at bay.

So if you want people to pay attention in virtual meetings, consider scheduling them in the afternoons, keeping them short, inviting fewer guests, and opting for Fridays when you can.

## ONLINE EMPATHY: THE VIRTUAL DISENGAGEMENT HYPOTHESIS

Imagine trying to explain social media to your great-great-great-great-grandparents.

You tell them that in today's world, almost two-thirds of the global population spends a whopping 2.5 hours a day in an online world where we can communicate with just about anyone. Based on this, they might envision a world united. To someone who has never used social media, this sounds like a gateway to world peace, with international harmony on a scale never before seen. After all, social networking apps have created a global marketplace for social exchange, offering unprecedented interconnectedness. Yet the reality is far from utopian.

Social networks resemble a hotbed of hostility, rife with hate speech and inconsiderate behavior. Studies reveal that for most people, the social media experience is anything but harmonious. In America, 52 percent of adults have been harassed online, and 51 percent of teens report being harassed on social media in just the last year. More than 1 in 5 Americans admit to getting in arguments "sometimes" or "often" on the internet. If you want to see this evidence for yourself, simply scroll through the comments on any viral post. I'm sure you'll find numerous harsh attacks and users bickering viciously with one another.

I've certainly experienced this myself. I get hateful comments all the time on my social media posts—it just comes with the territory. About 1 in every 15 interactions on social media contain negative sentiment, which might not sound so bad if you assume the rest are positive, but only 6.51 percent are. Hostility is an everyday occurrence, but a few years ago, I had a single encounter online that changed everything. I was working at Stanford University, leading a research project studying the neuroscience of empathy. I spent my days in the lab trying to figure out how empathy could be *enhanced*, while my evenings were spent navigating the often-hostile environment of social media. The disconnect was unsettling. I feel terrible for the many innocent people who have endured the unnecessary brutalization that virality seems to bring.

Then one day, a stranger posted a harsh video about me on TikTok. They were speaking to their camera and simply explaining what they didn't like about me. They argued that I wasn't trustworthy, and even implied that I was narcissistic. Naturally, I was hurt. Unsure how to respond, I left a comment. I suppose I just wanted them to know that I'd watched their video. Then everything changed.

The creator of the video replied with a series of comments, apologizing profusely. "I didn't think you would see this and I didn't mean to be insulting," they wrote. In another apology that they sent to me privately, the creator explained they had been projecting their stress onto me after a difficult week.

This was amazing to me. Suddenly it all seemed so obvious. The person behind the video wasn't a monster—they were actually quite empathetic. Otherwise, they wouldn't have issued such a genuine and heartfelt apology. It was almost like their brain hadn't perceived me as being a real person with real feelings until I left that comment. Social media had *temporarily suspended their empathy.*

This incident highlights the crucial role of *social cues* in empathy. Our brains are wired to detect tone of voice, facial expressions, and body language—all signals that tell the brain it's time to fire up those empathy systems. Indeed, those empathy-related brain areas light up on brain scans when people process facial expressions or emotional sounds like someone laughing or crying. This makes perfect sense considering that the brain's empathy systems evolved over millions of years of *face-to-face* social contact. However, those valuable social cues are almost completely eliminated on social media. Our complex emotions are reduced to plain text on a screen, humanized only by a small circular avatar displaying our image. Can one really feel empathy for a person reduced to this vague, inhuman representation?

I wasn't in the room while this person filmed their TikTok video, and they weren't in the room when I watched it. Without observing any of my social cues, how *could* their brain empathize with my feelings? Just as a tree that falls in an empty forest (disputably) doesn't make a sound, an emotion that is expressed with no bystander can't recruit any empathy. In comparison, imagine that you are in elementary school and you placed an upright thumbtack on the floor strategically to prank your teacher. Somehow, that tack managed to remain un-stepped-on until just forty-five minutes ago, when an innocent student stepped on it. As you read this now, this student is crying at the nurse's office, waiting for their parents to pick them up. Did you feel empathy forty-five minutes ago? Of course not, because you weren't there to *observe* their pain. However, you may feel it now, as you realize that you are responsible for an innocent child's suffering.

On social media, we are not present for the "stepping on the tack" part of the interaction. We leave insulting comments or create mean videos, triggering emotional responses that we will not witness. Interactions on social media are not *synchronous*.

Aside from facial expressions and vocal tone, there might be other unexpected factors left out in virtual interaction, like social smells. A recent study found that when men sniffed emotional tears from women, they not only *became less aggressive* but also showed reduced activity in aggression-related brain areas! This system may exist to help us relent when arguments get out of control and emotions rise to a peak. However, you certainly can't smell your enemy's tears on Facebook, so that aggression may continue unchecked.

Now, there's an argument to be made against what I'm proposing: that *videos* on social media permit users to see and hear one another's social cues, so empathy should be possible. There are two problems here. First, the videos we watch on social media are *replayed* rather than live, and although that may seem inconsequential, it makes a surprisingly big difference for the brain. In one study, subjects were shown a video of someone's eyes and asked to imagine what they were thinking or feeling. Little did they know, they were sometimes watching a live feed of the other person, while other times they were seeing a prerecorded video. Shockingly, the subjects showed reduced activity in empathy-related brain areas when watching the *replayed* video, even though they didn't know it was replayed. For some reason, watching replayed social content (as we do on social media) doesn't entice those empathy-related brain areas as strongly—and that may be underlying the hostility running rampant in these virtual spaces. Second, people experience less mimicry when watching prerecorded videos than when seeing live videos, and show less activity in brain areas containing mirror neurons. The replayed context of these videos thus may have a serious impact on how our brains engage (or fail to engage) with the person behind the post.

You could say that I was in the wrong place at the wrong time to receive this harsh, derogatory TikTok video. It could've been anyone, but I happened to be the unlucky guy that they took out their frustration on. I would argue the opposite. This incident crystallized a key idea and made it all click into place: Virtual interactions simply do not engage our empathy systems as strongly. It seems to be yet another of the brain's internal barriers that only divides us and creates chaos rather than peace. In 2024, I published this idea with my student Maria Tavares, naming it *the virtual disengagement hypothesis*, referring to the disengagement of empathy-related brain regions online.

Of course, low empathy isn't the only culprit responsible for our toxic online culture. There are other factors at play, like *anonymity*. People are more likely to *troll* others (purposely insult or upset them) when operating under an anonymous identity, and we all have the liberty to be anonymous online. When your identity is concealed, you can ruthlessly attack others without attaching it to your reputation, and unfortunately that's a powerful incentive to do the wrong thing.

Another factor is *perceived distance*. There's a real physical gap separating the harasser from their victim online. When you insult someone face-to-face, there's a lot more to lose—specifically, your teeth. However, virtual arguments open the door to penalty-free conflict. Just think of how much easier it is to leave a note on someone's car that reads, "You parked like a dick, you jackass!" versus waiting for them to return and saying it to their face. One of those options is clearly preferable, unless you enjoy getting your ass kicked.

While we can't rule out the influence of these factors, I really believe a lack of empathy is a central player. Studies indicate that the strongest predictor of online harassment is *psychopathy*—a condition characterized by a lack of empathy. This seems to confirm that empathy is a focal factor in our online problem.

Unfortunately, the consequences of this emotional disengagement can be devastating, as the victims of social media abuse feel the full impact. You know this firsthand if you've ever been harassed online—it sucks. Sadly, cyberbullied teens are 2.57 times more likely to attempt suicide. The bad actor can walk away without concern, as the victim's pain has no route to impress an emotional burden upon them. Meanwhile, the victims feel genuine anguish. It's as if the dynamic of online conflict is tilted, with all the weight falling on the recipient of jabs and insults. On one side of the phone, emotions are blunted, while they remain perfectly intact on the other.

What's more, the pain people experience from being attacked online is very real. Neuroscientists Dr. Naomi Eisenberger and Dr. Matthew Lieberman have shown that *social pain* activates many of the same brain areas as *physical pain*. Their work reveals that people who are socially excluded show patterns of brain activity highly similar to those accompanying physical pain. Lieberman details this study and many more in his 2013 book, *Social*. Perhaps most notably, these studies were conducted using *virtual* interactions. Eisenberger and Lieberman found this while subjects played Cyberball on a screen in a brain scanner (again, this was only done digitally because it would be impossible to play *real* catch in a brain scanner). Because of this, our fundamental knowledge about how the brain handles social pain comes from *virtual* interactions—and it shows that even these distant interactions can be truly painful.

As humanity deepens its investment in online communities, we must keep this in mind. Even though we can't see them, there are real people on the other side of our attacks, with real emotions. So next time you're inclined to treat someone hurtfully online, be aware that the pain you are causing feels very real to them, and it really hurts.

⬭

It may seem like the internet's harshness is something new—perhaps a recent by-product of the growth of social media—but the truth is, people have been acting weird online ever since the birth of the internet. Back in 1984, scientists at Carnegie Mellon University reported that people interacting through this new thing called a computer were more prone to "swearing, insults, name calling, and hostile comments." It seems that social media isn't to blame. Rather, the human brain might just be a jerk when it can't see the person it's interacting with. Interestingly, mice do this, too. Remember in the last chapter we learned about how mice will (adorably) choose to be generous and share treats with other mice? In those experiments, the recipient mouse was on the other side of a mesh grid wall, allowing the generous mouse to smell, hear, and even sort of touch them. But when the researchers replaced that mesh grid with a solid divider to prevent the mice from interacting or detecting one another's social cues, those generous behaviors plummeted. As they say, "Out of sight, out of mind."

But if it feels like it's been getting worse, it is, at least in the United States. Between 2014 and 2020, the share of Americans who received physical threats on social media doubled from 7 percent to 14 percent. Name-calling and purposeful embarrassment also rose during this period.

I can't help but notice that this coincides with another major shift in the United States: an extreme burst in political polarization. Over the last decade or more, our country has grown increasingly divided over politics. The political affiliations of our coworkers, neighbors, and even family and friends have become much more relevant, exposing yet another way to divide *us* from *them*. We've begun demonizing those on the other side of the political aisle, viewing them as members of the opposing team. That's bad, considering that the brain exhibits less empathy for those in our outgroup. Political polarization is the perfect storm for a plunge in empathy, and I believe that our sticky, hostile political climate has oozed its way into online spaces, making the internet's already-bad empathy issues even worse.

If we wish to build a healthier online world, the way forward is remarkably simple: Exercise the same pro-empathy tips that we would in person. When you feel the impulse to leave a not-so-charitable comment on someone's post, pause for a moment to consider how your words would impact them. Engage cognitive empathy to understand the headspace that your comment would place them in. Then try to embody those emotions through emotional empathy. If it helps you step into those emotions, imagine someone you care deeply about reading that comment, like a parent or child. Would they be hurt? If so, can you make your point in a more positive and constructive way? Remember, although you may not be there to witness them "step on the tack," there is always a real person on the other end of your message, and those tacks really sting.

When it comes to online fights, there's only one certainty I can promise: You will never convince someone they're wrong. If someone is willing to invest time and energy in arguing with a stranger on the internet over this topic, there's probably very little flexibility in their viewpoint. You're not going to be the one to change their mind. Kindness and respect will always get you further than anger and derision. Plus, engaging in an argument will only ruin your day. Nothing keeps you glued to your phone like angrily awaiting the reply of a stranger. After each response, you'll obsess over how your last comment could've been better and plan your next diabolical retort. Of all the amazing things humans can do with our remarkable brains, arguing online is probably one of the most useless ones. I think arguing on the internet is good for nobody, and that opinion is informed by science: Online arguments generate more negative emotions than in-person disputes and are less likely to result in consensus. You're really not getting anywhere. Ignore the trolls, and avoid becoming one yourself.

## CAN EMOJIS SAVE US?

Let's face it, this chapter has been pretty gloomy. I've been quite a Debbie Downer the way I've bashed social media and virtual interactions. It seems very depressing, but there must be some good to speak of . . . right? Some glimmer of hope?

Fortunately, an unlikely hero may be able to redeem us from the perils of the internet. A savior that hides in secrecy behind a flippant yellow mask: *emojis.*

I know, I know. You're kidding me, right? It sounds nuts . . . but hear me out. Recall that as we move down the line all the way from real-life interactions to text messages, we gradually strip away important social cues that promote things like empathy. Well, what if we could restore facial expressions by using emojis? Perhaps this would make things more lifelike and engage the social brain a bit more. If we're lucky, maybe it can even sprinkle a touch more empathy into our text-based interactions. Amazingly, some research suggests that this may be possible. Emojis expressing pain (like 😖 or 😣) have been shown to trigger an empathy-like response from the human brain!

As you know, pain is very visible on the human face. For a moment, close your eyes and picture someone in pain. Most likely, you imagined one of two things: either their mouth and eyes wide open in a cartoony sort of "YOWW!!" expression of pain, or their eyes shut, brows furrowed, and teeth clenched in a "Damn it, I just stubbed my toe!" kind of expression. Let's focus on this second one. If we strapped you into an EEG to measure the electrical activity from the surface of your brain, we would see a predictable brain response when you look at this type of facial expression. The pattern we see would look slightly different than when you look at a neutral face that is not expressing pain, and that difference is thought to represent part of the brain's empathy response. Shockingly, EEG recordings reveal that people show this same pattern when looking at an *emoji* in pain! Even though it's an unrealistic, cartoony face (like this: ☹), it evokes a response similar to a human expression of pain.

Of course the brain's reaction to an emoji face isn't *exactly* identical, and there's a simple reason why: Processing human faces is a much more complex task for the brain. Naturally there's a lot more information to take in. However, the fact that there's any overlap at all suggests that emojis *do* represent some element of the corresponding human emotion, and the brain respects them accordingly.

I believe this could be a useful tool for enriching our text-based interactions. Emails, texts, and comments on social media leave out the social cues that typically frame our words within the appropriate emotions and convey our internal state. This free-standing text may ring hollow and cold without those supportive cues. Maybe emojis can help restore some of that information, assigning the appropriate emotional valence to our messages. For example, compare the following text conversation. Which messages do you feel are more heartfelt, genuine, and impactful?

**Message**

A. I heard you were hospitalized. I hope you're okay.

B. I heard you were hospitalized. 🥺 I hope you're okay. 🙏

### Response

A. Thank you for your message. I am feeling better.

B. Thank you for your message. ☺ I am feeling better. 😄

If you're like me, you felt the emotions more strongly in the messages with emojis. They felt deeper, even though the content was identical. Perhaps this is because the emojis triggered the neural responses that the corresponding facial expressions would evoke. However, your feelings about emojis likely play a huge role here. If you view them as childish and insincere, these messages probably just felt corny.

This is an emerging science, and frankly, what I'm proposing here could be a total reach. For right now, it seems plausible that emojis are a legitimate way to enrich and deepen virtual connection. I might end up being proven totally wrong, in which case I will publicly declare that I am stupid. You have my word on that—I'll call a press conference. But for now, it certainly can't hurt to throw an emoji in there. Always worth a try, right? 😉

### Key Takeaways

1. Humankind has been barreling toward nonsynchronous interactions for centuries. However, the brain is built on millions of years of face-to-face contact, making virtual interactions highly abnormal for the social brain.

2. Our bodies are naturally equipped with social indicators. When these cues are stripped away through phone calls or texts, the quality of our interactions suffers.

3. More lifelike interactions provide greater benefits to mood and well-being. Thus, interacting over video is preferable to texting, but in person is best.

4. Using phones during in-person interactions (phubbing) compromises the value of the interaction.

5. The effects of social media are the opposite of those of socializing. Social media use leads to lessened well-being and worsened loneliness.

6. People may experience less empathy online due to the lack of detectable social cues. This could be why virtual spaces tend to bring out excessive levels of hostility and anger.

7. Emojis evoke brain responses similar to those triggered by human faces and could be helpful for enriching text-based interactions.

To view the references cited in this chapter, please visit benrein .com/book.

# 7

## TANGLED WIRES

How Love, Touch, and Deep
Connections Tickle the Brain

This year, about 150,000 Americans will be diagnosed with colorectal cancer. Sadly, roughly a third of them will die.

The survivors will be somewhat predictable. If you were handed a long list of these 150,000 people and their information—age, weight, diet, habits, etc.—you could probably pick them out above chance level. Statistically speaking, they will be younger and healthier. They will be the people who exercise, eat a healthy diet, and don't smoke. They'll be the ones who opt for leading treatments like surgery and chemotherapy. And perhaps unexpectedly, they will be the patients who are *married*.

At least that's what a 2013 study suggests, which tracked over seven hundred thousand American cancer patients and compared their outcomes by relationship status. What they found was remarkable. For those with colorectal cancer, being married reduced the likelihood of death by 28 percent. The same trend was found in other forms of cancer, like prostate cancer (24 percent), breast cancer (22 percent), esophageal cancer (23 percent), and head/neck cancers (33 percent). When the researchers compared these numbers with the existing data on chemotherapy, they revealed the most incredible finding of all. In five out of the nine cancers they examined, being married had a stronger influence on someone's survival chances than receiving chemotherapy did.

That doesn't mean chemotherapy isn't effective—of course it is. Rather, the protective effects of marriage are probably related to chemotherapy, as spouses often encourage each other to get the best possible treatment. Indeed, the study found that unmarried patients were often undertreated for their cancer, perhaps because they didn't have a partner encouraging them.

Spouses offer more than just good advice, though. They also provide precious *social support*. Even when we're not facing the challenging reality of something like cancer, there is immense value in coming home from stressful days to the care of your lifelong partner. This social support can make all the difference, and in cancer patients, it has been shown to help buffer against stress and preserve well-being.

But there's more to it. These powerful lifelong relationships also carry special benefits for the brain and body. They prop up our well-being in exaggerated ways, exceeding what we get from talking to a stranger on the train. They also usher in opportunities for parenthood, which can offer similarly potent therapeutic benefits. Both romantic relationships and parent/child bonds are unlike other connections; they stand a level above. There is a defiant glue that holds them together, unbreakable by typical forces, and with that comes a slew of distinctly favorable neural properties.

Just like some foods may exceed the nutritional offerings of others, or how a night in your own bed may leave you feeling more rested than one on an air mattress, is it possible that certain relationships may offer disproportionate benefits for your brain? In a divided world, where we must seek unity for the sake of our brains and wellness, I worry we may be overlooking the rich value of certain special bonds. To appreciate the full picture of our modern social problem, we must wrap our heads around the deepest, most powerful connections of all, and what they offer the brain. It's time for us examine the neuroscience of love, and all that comes with it.

## LOVE ON THE BRAIN

Have you ever been "truly, deeply, and madly in love"? If so, I'm happy for you. Also, you'd be the perfect research subject for Drs. Andreas Bartels and Semir Zeki.

In the late 1990s, the pair of scientists posted an advertisement containing this exact phrase, hoping to recruit some lovebugs. The subjects they enlisted were asked to sit in a brain scanner while they thought about their partner. As you might guess, the researchers were hoping to uncover the neuroscience of love. Fortunately, they seemed to find it. When people thought about their beloved, they showed a unique pattern of brain activity. This pattern differed from when they thought about a friend who was the same sex as their partner. Dreaming of their lover activated brain areas like the insula and anterior cingulate cortex—those same regions that drive emotional empathy—along with the ventral tegmental area, globus pallidus, nucleus basalis of Meynert, and the substantia nigra. While these last four areas might sound like nothing more than a magical incantation, a common theme ties them together. **They all have oxytocin receptors.**

Oxytocin is the MVP of social bonding. Recall from chapter 1 that it plays a key role in *social reward* by driving the release of serotonin and dopamine. Can you think of a time when you've felt extremely rewarded by someone's presence? Has there ever been someone you never got sick of and always wanted to be around? If you've ever fallen in love, this might sound familiar.

The early stages of a relationship are often enchanting. Everything about your new partner feels *perfect*, and you hardly want to leave their side. This feeling may be explained by a heavy presence of oxytocin, which creates a profound sense of social reward. Humans, like all animals, have two primitive instincts: to survive and to reproduce. So when you meet a potential mate, your brain releases oxytocin to make you feel *rewarded* for being around them. Their presence feels reinforcing, so that ideally you fall in love, make babies, and humans continue to thrive.

What I mean to say is, when brains fall in love, you can bet your bottom dollar that oxytocin is flowing. Israeli scientist Dr. Ruth Feldman measured oxytocin levels in newly minted lovers (those just a few months into a new relationship) and found that their oxytocin levels were roughly *twice* as high as those of singles. This probably explains why Bartels and Zeki saw activity in those brain regions that express oxytocin receptors. If oxytocin is the glue that holds together romantic relationships, it should be doing its thing when people think about their partner, thereby activating those brain areas.

Six months after Dr. Feldman's team took those oxytocin measurements, they invited back several of those couples to get some more samples. Now that the honeymoon phase was over, did their oxytocin levels drop? Interestingly, they did not. Even though half a year had passed, the oxytocin was still flowing at about the same levels. This suggests that oxytocin doesn't just help glue people together at the beginning—it also provides sustained support over time.

Now, what happens when you try sticking two things together, but you don't use enough glue? Naturally, they separate. This seems to be true in relationships, too. In Dr. Feldman's studies, many of the couples sadly didn't stick together (pun intended) until the six-month mark. The researchers cleverly saw this as an opportunity. They looked back at their original oxytocin measurements and found that the couples who didn't last had shown *significantly lower oxytocin levels* early in the relationship. With less oxytocin flowing, those couples may have found each other less rewarding and consequently failed to form a resilient connection. By simply looking at people's oxytocin levels, science may be able to predict the outcome of their relationship.

Based on all this, you could make a strong argument that what we call love is just *an abundance of social reward*. Love might just be what happens when our brains produce heaps and oodles of oxytocin, making us feel rewarded by someone's presence and satisfaction from spending time together. If love is just the brain's social reward systems being *maxed out,* then we might expect love and oxytocin to be functionally synonymous—and that is basically what we've seen here.

But what about long-term relationships? When couples stick together for *decades,* does oxytocin remain high or gradually decline? One study answered this by scanning the brains of happily married couples—together for twenty-one years on average—while they thought about their spouse. The project was positioned to answer a heart-wrenching question: When the passion and excitement of the honeymoon phase is long gone, do lovers still light up each other's brains, or have those oxytocin signals faded?

Fortunately, they found what we'd all hope for. The brains of the long-term partners showed very similar patterns as those of the fresh lovers. Almost all the same brain areas lit up, including those with oxytocin receptors. It seems that even after decades of enduring love, romantic partners still light up the brain in the same magical way. (Who's cutting onions again?!)

Sorry to kill the cutesy vibe here, but I have to add that this doesn't mean *all* long-term relationships light up the brain's oxytocin systems. Those long-term couples were *happily* married; even after all that time, they maintained that they were still "madly in love" with their partner. Research shows that married couples who report greater relationship quality show higher oxytocin levels. This all suggests—uncontroversially—that being happily in love is associated with higher oxytocin levels. And I believe that's something that we should be paying *much* more attention to.

That's not just because love is pleasant and wonderful. It's because oxytocin does *so much more* than just help us form bonds. It also does loads of extraordinarily healthy things within our brains and bodies, to the point that it's even been called "nature's medicine" by some researchers. First, animal studies suggest that **oxytocin can suppress stress and anxiety**. For instance, when researchers deleted the oxytocin gene from female mice, the mice behaved much more anxiously. If given some oxytocin, the mice seemed to relax. In humans, too, oxytocin may help manage social stress. One study gave fifty-four singers oxytocin or placebo before they performed in a public setting, and those who got oxytocin felt that their performance went much better.

Second, **oxytocin is anti-inflammatory**. When mice are injected with a compound that causes an inflammatory response in the brain, their being pretreated with oxytocin reduces the levels of inflammation it triggers. Remarkably, this is also true in humans: When men are given oxytocin along with that same compound, they show a reduced inflammatory response.

Third, **oxytocin is neuroprotective**. When mice are given oxytocin after a stroke, they show a smaller area of damage in the brain and less neuroinflammation. Similarly, human stroke patients with stronger social support show greater recovery after stroke.

Fourth, **oxytocin supports wound healing**. Wounds generally take longer to heal in those who are stressed, but when stressed hamsters were given oxytocin, their wounds healed faster. The same effect was found when they were allowed to hang out with other hamsters, presumably thanks to the natural oxytocin release it causes. Higher oxytocin levels have been associated with faster wound healing in humans, too.

Now, let's just pause for a moment. Isn't this remarkable? I find it astounding how many systems throughout the body oxytocin supports—and that's not even all of it. Oxytocin has also been shown to strengthen the immune system and support the buildup of bones (did you know bone cells express oxytocin receptors?). But why is this the case? Why does oxytocin have all these wonderful effects—like reducing inflammation—that have nothing to do with social bonding?

Think back for a moment to what we learned about cortisol in chapter 2, and its role in stress. In the short term, cortisol reduces inflammation to help prepare us for whatever challenge or battle we may face. Oxytocin may be anti-inflammatory for similar reasons: to promote reproductive health. When we meet a potential mate, it's vital that we're able to reproduce, for the sake of keeping our species around. Oxytocin may prepare us for that responsibility by lowering inflammation and promoting overall wellness.

Now remember what we just learned: People in happy, loving relationships have higher oxytocin levels. When they think about their partner, brain areas that express oxytocin receptors light up. This is a powerful declaration. It means that the love we receive from romantic partners isn't just a source of joy and happiness— it's a fountain from which "nature's medicine" can flow. Thanks to the restorative nature of oxytocin, love may offer amazing benefits for our health and well-being. This is yet another gift of biology and evolution. It's as if Mother Nature curated the human experience around finding love and building a family, rewarding those who do with tremendous protection. I believe this is something we should take advantage of. And now, thinking back to those 150,000 cancer patients, **is it really any wonder that the married ones will be more likely to survive?**

## WHY BRAINS LIKE SEX

One of the most potent features of romantic relationships is the intimacy they offer. To really appreciate what makes these bonds unique, we must look at the brain signature of sex. Although it may come as a surprise, there's quite a bit of research on this topic!

Believe it or not, multiple studies have crammed people into brain scanners while they either masturbated or received genital stimulation from their partner. Of course, to have sex in a brain scanner would be nearly impossible. Or at least it wouldn't be very good.

What these studies reveal is—drumroll please—a whole lot of brain activity. During sexual arousal (the buildup period leading up to orgasm), brain activity gradually rises across multiple regions, like a symphony growing steadily louder. Then brain activity peaks during orgasm—reaching a noisy crescendo—before returning to its typical hum. Interestingly, a different set of brain areas come online during orgasm as opposed to that arousal period. The symphony that activates during orgasm isn't totally figured out yet, because as you can imagine (or maybe don't), it's challenging to study orgasm in a brain scanner. That's because orgasms are short and fleeting; often involve some amount of movement, which can blur brain scans; and are difficult to achieve in the presence of scientists. However, it has been done, and here's what has been found:

One of the regions activated during an orgasm is the ventral tegmental area—the brain's largest source of dopamine. This is no surprise if we consider dopamine's essential role in motivation and reward. Because of how crucial reproduction is for the survival of humans and other species, evolution made sure that orgasms were intensely rewarding. Studies in rats suggest that oxytocin also plays a role, finding that oxytocin-producing neurons were more active after sex. It's noteworthy that orgasm seems to involve dopamine and oxytocin, two of the three players that drive social reward. Any idea why that might be?

Well, biologically speaking, when two people experience an orgasm together, there is a chance that a baby will soon be on the way. That baby is more likely to survive if the parents work together to raise and protect it. So having overlapping brain systems for orgasm and social reward may increase the likelihood that two people *bond* through sex. It may be evolution's way of injecting a bit of that neural glue between sexual partners, for the sake of the baby's survival. That way, we don't just feel rewarded by the orgasm itself, but by the presence of the person we shared it with.

But what about the intensely euphoric experience of orgasm?

There must be some other system in the brain behind this. If just oxytocin and dopamine were involved, then hanging out with friends might also feel like an orgasm, but luckily that's not the case.

For those intense euphoric feelings, we may owe thanks to endogenous[+] opioids (aka endorphins). These are the brain's natural painkillers, and research has shown that when men received a hand job from their partner (sorry, no scientific term for that one), they showed increased opioid signaling in the hippocampus. Meanwhile, another study found that giving men Naltrexone—a drug that blocks opioid receptors—increased the intensity of orgasms. While it's still unclear what exact role endorphins play in orgasm, they seem to be involved somehow. Teasing this apart will be on the frontier of orgasm neuroscience.

I don't think anyone really needs a neuroscience book to tell them that sex is good and feels nice, but I do think it's something worth looking at in a modern society that's rapidly plunging toward digitalization of everything. Many forces are pulling us apart, and we're now contesting a new one: AI-powered large language models like ChatGPT.

Could ChatGPT possibly pose a threat to the sexual prowess of humanity? Alas, there's reason for concern. A 2024 report analyzed almost two hundred thousand conversations with AI chatbots and found that 7 percent involved sexual content. Sure, 7 percent might not seem like a lot . . . but it sort of is, isn't it?

Look, I wasn't in the room with these people during these horny robot conversations (thank goodness), and I can't claim to know what they were up to. All I'm saying is, we should be cautious about turning to non-intimate, AI-powered alternatives to human connection. Call me old-school, but I urge humanity to not go that route. Not only would this keep us from reaping the benefits of social and sexual contact, but it also may be fruitless. Because as you'll learn shortly, our brains simply don't respond the same way to intimacy from a nonhuman entity.

⊙⊙

Another biological gift offered by romantic relationships is *physical touch*, the importance of which cannot be overlooked. Research shows that various forms of touch (like massages or hugs) can help regulate cortisol levels, reduce pain, lower depression scores, and reduce anxiety in adults. Massage has also been linked to oxytocin release, which we know can have many great benefits. For the average adult, most of this healing touch will come from a romantic partner; however, it's worth noting that the benefits appear consistent regardless of who's doing the touching. Research shows that touch provides mental and physical benefits regardless of whether it's coming from a family member, friend, partner, or health care professional.

In order to appreciate *why* touch has such great benefits, we should first look at *how* the brain processes touch. Let's start in the skin. Your body is densely packed with touch receptors—also known as *mechanoreceptors*—which come in a variety of flavors. Each type responds to a specific form of touch. Some are activated by vibration or light pressure, while others might respond to stretching of the skin. That's because it's important to know whether that's a snake sliding next to your ankle or just the cuff of your pants blowing in the wind. Having different types of touch receptors allows for precise detection of sensations on the skin. One sensor in particular—called *C tactile fibers*—detects slow, light, caressing touch. These bad boys are thought to be responsible for processing *pleasant* touch. Now let's follow these fibers back to the brain.

After snaking their way up the spinal cord and under the skull, C tactile fibers activate two brain areas associated with reward processing. The first is the orbitofrontal cortex, which we'll call the OFC. This region is located at the front and bottom of the brain, just above your eyes. In fact, that's where the name comes from. It's just above the cavities called *orbits* that contain the eyes. The level of activity in the OFC is related to how rewarding something is. For example, imagine you sit down with a delicious-looking plate of steak and mashed potatoes. You dive into the potatoes, devouring bite after bite. After you take a few mouthfuls, they start to lose their luster as you become less interested in eating them. Meanwhile, you still haven't tried the steak and it's looking *good*. In this moment, your OFC will probably be more active when you taste or even smell the steak compared to the potatoes, because the steak has more reward value to you. The OFC helps us make decisions based on how rewarding our options are. In this case, it would likely suggest that you take a bite of steak next, instead of potatoes. Now remember, when we experience gentle touch, those C tactile fibers activate the OFC, suggesting that the brain perceives that touch as pleasant and rewarding.

The second brain area activated by pleasant touch is the posterior superior temporal sulcus (or pSTS), which runs along both sides of the brain just above the ears. When people are gently touched, they show more activity in the pSTS when they rate the sensation as more positive. If you've ever delighted in the sensation of a back massage or foot rub, you shouldn't just be thanking the masseuse, but also your pSTS and OFC.

But here's the thing: It's crucial that it's a *someone* rubbing our back, and not a *something*. Because as I mentioned earlier, the brain responds differently when the masseuse is nonhuman. The evidence for this comes from a study where people received a foot massage from either a human or a robot. Even though they were blindfolded so they didn't know which was massaging them, they showed less activity in the pSTS, rated it as less pleasurable, and showed **lower levels of oxytocin release** when the massage was coming from a robot.

What does this mean? Why did they produce less oxytocin during the robot massage? Did the robot just give a worse massage, or was this a sign of something more profound? It may sound outlandish, but what if their bodies could somehow detect that they were being massaged by a human, and were releasing oxytocin to *form a bond* with the person behind the hands? This could explain why less was released during the robot massage: because there was no person to bond with.

Indeed, a year later, the same researchers published a follow-up study that explained everything. Once again, the subjects were massaged by a human or a robot, but this time they got some oxytocin beforehand. The result? Oxytocin made the massage more pleasant, but *only when they were being massaged by a human*. Oxytocin also increased brain activity in the OFC and STS *only* during the human massage. What the hell was going on here?

As you've now heard a million times, oxytocin drives social reward. So when oxytocin is released during a massage, it may make the sensation more pleasurable. As a result, this could strengthen your social bond with the masseuse. After all, when someone touches you in a way that feels good, you feel connected with them, right? But if the masseuse is a robot, there's no use. Most people don't want to form a bond with a massage machine (although I have formed bonds with massage chairs at the mall while my wife was shopping). This could be why oxytocin *didn't* enhance the pleasure of the robot-delivered massage: because there was no living, breathing being on the other side to form a bond with. In other words, the brain may release oxytocin to enhance the massage only when we *want* to bond with the person delivering it.

One study provides clear evidence for this. A group of heterosexual men (their sexuality will become relevant in a moment) were introduced to a male and a female research assistant, then given some oxytocin and sent into the tube of a brain scanner. As they laid there, a piece of fabric was draped over the opening of the scanner so they couldn't see their lower body. Then someone entered the room and started gently touching their legs. In some cases, the men were told it was the male experimenter they'd just met, while in other cases they were told it was the female. Fascinatingly, the men who were given oxytocin rated the touch as more pleasant only when they *thought it was the woman*. They also showed increased activity in the OFC. This is where their sexuality comes in. Being heterosexual, the men would be keener to form a bond with the female experimenter than with the male. Adding oxytocin to their system may have accelerated their brain's drive to affiliate with the woman, and it did so by selectively making *her* touch feel more pleasant.

Years later, the researchers replicated this experiment with heterosexual *couples* instead of individual men. Some of the subjects were told that the mystery masseuse behind the curtain was their partner, while others were told it was a stranger of the opposite sex. Once again, those who got oxytocin beforehand rated the touch as more pleasant only when they thought it was coming from their partner, and not the stranger. This suggests that oxytocin intertwines itself with someone's touch—and perhaps makes it feel better—to support bonding with those we want to become closer with.

This may have profound implications for the health benefits of romantic touch. When our partners massage us, we might not only find it more pleasant than when other people touch us but also *release more oxytocin*. Considering the powerful benefits of oxytocin, this could be huge for our health. It's another no-cost intervention that could have a positive influence on our well-being. You may have the rare ability to evoke higher levels of pleasure, reward-related brain activity, and oxytocin in your partner, so just go touch them. Maybe cuddle up to them or offer a massage. Tell them your book made you do it.

## THANKS, MOM

Few social bonds are as neurologically potent as romantic love, but there is one that comes close: the bond between parent and child. Just like romantic bonds, these relationships are lathered in oxytocin. Seriously, oxytocin levels surge when parents do damn near *anything* with their child. When parents play with their child or make physical contact, oxytocin levels go up in both the parent and child. When parents simply *look* at their child, oxytocin levels go up. Hell, a mother's oxytocin will even rise when their baby starts *crying*! This is all for a good reason: These copious amounts of oxytocin make it enormously rewarding and delightful for parents to be with their children. By making parent/child interactions so reinforcing, parents are motivated to care for their young. After all, if baby's parents don't take care of baby, who will take care of baby? Meanwhile, the infant also needs to be motivated to attach to its parents for protection and care. Evolution needed a brain system to glue parents and children together, and oxytocin was the answer.

Oxytocin's role in the parent/child relationship begins sooner than you might expect, though; in fact it starts even before the two have a chance to meet face-to-face. Surprisingly, oxytocin takes on a special responsibility while the baby is still in the womb. Or to be more precise, *on the way out of the womb.*

Mothers produce large amounts of oxytocin during childbirth. If you were to speculate why, you might assume that's because the mother's brain is anticipating the baby's arrival and is getting ready to form a strong bond. However, it turns out this oxytocin may have a very different purpose: *to protect the infant's brain during birth.*

You don't remember experiencing it yourself, but childbirth isn't gentle. It's a positively turbulent and stressful event. On the way out, the baby is liable to face periods of restricted blood flow that could limit how much oxygen reaches the baby's brain. That, of course, would be very bad. Luckily, the oxytocin produced by the mother is thought to transfer to the baby and *protect* its neurons against suffocation. In the baby's brain, this oxytocin triggers a shift in how neurons signal that lowers their metabolic demands. In other words, it reduces how much energy the neurons are using. Think of it this way: If you're stuck underwater and you're running out of air, you will be able to stay conscious longer if you relax your body and limit your energy expenditure. Similarly, this oxytocin allows the baby's brain cells to metaphorically "hold their breath" for longer and survive periods of limited oxygen. Amazingly, it seems that before a mother even holds her newborn, her body is shielding them from harm.

This reminds me of a story my parents have always told about my birth. On my way into the world, I faced my very first challenge in life: The umbilical cord became wrapped around my neck. The doctor noticed that my heart rate was dropping dramatically with each contraction, and she immediately ordered an emergency C-section. Yes, I made it out okay—but I look back on this event now with wonder and curiosity. In those dire moments while I was grappling with strangulation, I wonder if my mom's oxytocin was flowing through the very same cord enclosing my throat to protect my young neurons. There's no way of knowing, but her oxytocin may have saved my life before it began—entering through the same conduit that threatened my existence.

While writing this book, I've spent a lot of time reflecting on the miracle of biology, marveling at how clever these systems are. I'm uncertain that even the world's smartest think tanks could come up with such a brilliant design. I'm grateful for the intrinsic genius of our bodies, which operate without a hint of active thought from our conscious minds. Thanks to these complex programs ingrained within us, we're all born under the protection of biology's ancient wisdom.

I think we all owe our moms a thank-you. Before your mother even held you for the first time, her body was protecting you and making sure you arrived safely. If you have the privilege of doing so, call your mom today. Thank her for the oxytocin.

⚭

I mentioned earlier that parent/child bonds are the only relationship that can rival the neural potency of romantic love. But how similar do they look in the brain? If you had to guess, would you estimate that there's a significant overlap in brain activity between the love one feels for a child and a romantic partner? Or do you surmise that these are two very different processes?

In 2004, Bartels and Zeki (the pair who first examined romantic love in the brain) struck again with another study. This time they examined mothers' brains while they thought about their infant. Compared to when mothers thought about a different child, thinking about their own baby uniquely activated certain brain areas like the insula, cingulate cortex, striatum, and substantia nigra. Notably, these brain areas overlap directly with activity seen in romantic lovers. It's not known for sure, but this group of brain areas may represent a sort of "love network" in the brain, which is core to infatuation for both children and lovers. If you guessed there would be overlap, you were correct.

However, you also were right if you predicted differences. When the mothers thought of their children, certain brain regions activated that *didn't* rev up during the scans of romantic lovers. The mothers showed unique activity in the OFC (yes, the same region activated by pleasant touch), and the periaqueductal gray: a small area deep in the brain stem that's involved in pain relief, threat detection, and spirituality. The activity in these areas may reflect the inimitable experience of loving a child, differing in certain ways from the love felt for romantic partners.

## TANGLED WIRES

I find it funny that in the English language, there's a phrase to describe when the thinking of two people is aligned: They're "on the same wavelength." What makes this amusing is how oddly fitting it is from a neuroscience perspective. That's because in research, an extraordinary phenomenon has been documented called *interbrain synchrony*, where the brains of two people interacting can somehow become synchronized and show nearly identical patterns of activity.

Interbrain synchrony is such a sci-fi type of thing that it's easy to misunderstand. You might think brains have some sort of frequency—like a radio broadcast—and if another brain gets close enough, the two brains snap into the same wavelength. That's *not* how it works at all. If that were the case, then interbrain synchrony would happen when you simply pass a neighbor on a narrow sidewalk or sit next to them on a bus. But no; interbrain synchrony typically happens only when two people are either *having a shared experience* or *cooperating*. This synchrony doesn't happen across the whole brain. Rather, it usually involves a certain brain area syncing up between two people, and the brain region differs depending on what they are doing.

When you pass that neighbor on the sidewalk, if you make eye contact and quietly cooperate to pass each other without colliding, then you may experience interbrain synchrony. That's because you're working together to achieve a common goal. Some cool neuroscience research has found that pianists show interbrain synchrony when they play a duet together for the same reason. (By the way, there's a fun name for measuring the activity of two brains at the same time: *hyperscanning*! It sounds super futuristic, or like an Olympic sport that involves copy machines.) So what's the point of having brains sync up? Well, research suggests that it may facilitate teamwork. When people work together on things like brainstorming or problem-solving, the teams with greater interbrain synchrony perform better. If you failed to achieve interbrain synchrony with your neighbor on the sidewalk, you might be more likely to bump into each other awkwardly, or land your foot in a puddle.

The other thing that triggers interbrain synchrony is shared experience. If you've ever felt the magic of being in a crowd at a concert, then you know what I mean. When everyone immerses themselves into the music emotionally, it's an amazing experience. In those moments, when you feel somehow *in tune* with a crowd of thousands of people, your brains may truly be syncing up. Research shows that concert attendees show interbrain synchrony only in certain moments: when everyone is collectively experiencing a high level of musical pleasure. During those electrifying flashes of musical fervor, interbrain synchrony peaks in the crowd. This makes me wonder how much interbrain synchrony happens every night, all over the world, in crowds attending musical events. If it truly peaks during the most exciting and pleasurable moments, I can't even imagine the level of synchrony when the bass drops at an electronic music festival.

Now, you're probably wondering why we abruptly jumped from talking about parent/child bonds to neighbors on the sidewalk and techno concerts. The reason is that the deep relationships we've focused on in this chapter tend to be "on the same wavelength" quite a lot. Romantic couples show much more interbrain synchrony than strangers do, and children synchronize more with their parents than with a stranger while playing a cooperative game. This privileged synchrony between parents and children can begin as early as infancy. Even before their first birthday, babies show stronger synchrony with their mother than with another woman.

Interbrain synchrony is highly dynamic—it *peaks* in couples during periods of eye contact and *decreases* between parents and children when they're competing *against* one another rather than cooperating. While these small details may seem trivial, they offer a glimpse into the true nature of interbrain synchrony: It's a cooperative pairing of brains, something that cannot be achieved between two enemies. It's biology saying, "You're on my team, so let's work together." However, it can easily be taken away. When loving parents and children are pitted against one another, the competitive nature of the interaction takes precedence and blocks interbrain synchrony. Apparently, the brain recognizes *allegiance* even over blood and sets its priorities accordingly.

I believe we should honor and take advantage of this incredible ability. It seems to improve our teamwork and collaboration, and that's significant. While most any two brains can achieve this synchrony, it's much more likely to happen between two brains that share a deep connection. This suggests we may be more effective at collaborating with those we are closest to, and some re-search suggests that may be thanks to our familiar friend oxytocin. When two men are given oxytocin before collaborating on a task, they show greater interbrain synchrony. Considering that oxytocin is extremely prevalent in (happy) couples and parent/ child duos, it's possible that this chemical is also responsible for the increased accessibility of interbrain synchrony.

As usual, this all makes sense from an evolutionary perspective. Interbrain synchrony likely evolved as a means for increasing our chances at survival by facilitating teamwork with those close to us. In a dangerous ancient world, it may have been crucial to cooperate best with your partner and children, as it's their lives you're most worried about protecting. The presence of oxytocin may assist this process, guiding two brains into the same headspace to cooperate and defeat a shared enemy or overcome other obstacles.

However, oxytocin can't be the only responsible party. There are surely other factors driving synchrony . . . like body odor. No, seriously. Recall that babies synchronize more with their mother than another woman? Well, when a T-shirt with the mother's scent was placed in front of the strange woman, the baby's brain synced up with her brain as if she were the baby's mother. At present, it's unknown whether scents continue to trigger synchrony in adults. They may just help babies recognize their mothers early in life while their vision is still developing. However, at the very least, it shows that lots of things likely drive interbrain synchrony. As usual, there are many secrets buried within our bodies that are shocking and impressive to learn about.

Sometimes science reveals things you'd never expect.

○○

As you've read this chapter, you may have wondered about another treasured type of bond: *friendships*. Does your best friend's brain sync up with yours, too?

Indeed, friends are more likely to show interbrain synchrony than strangers are . . . but only in the right circumstances. For instance, they're very likely to sync up while holding eye contact or working together on certain collaborative tasks. But in other situations, like cooperating to draw a picture together on an Etch-A-Sketch, friends show shockingly little synchrony—in fact, hardly any more than two strangers. It seems that while friends *are* more prone to synchrony than the average pairing of strangers, the link is not as robust as that in couples. In all these situations and more, romantic lovers showed much greater synchrony than friends did. In other words, interbrain synchrony is more accessible in friends than in strangers but is the most common in lovers.

This fits perfectly with our evolutionary explanation. If the primary goal of humans is to survive and reproduce, then we should synchronize most with our romantic partners and our children—the two key pillars of reproduction. Therefore, this rare ability shows up the most in these special relationships, and to a lesser extent between two friends (but still more than strangers).

While this may be kind of disappointing, here's the silver lining: You and your friends likely *do* share something else special—a similar brain. While your brains may not *synchronize* as readily as those of couples do, they might literally *look alike*. Research shows that friends show structural similarity in "social" brain regions like parts of the prefrontal cortex, temporal cortex, OFC, and amygdala. That is, the anatomy of the brains of friends is literally more similar than that of non-friends. It's not known for sure why, but one theory is that it comes down to the simple principle of *homophily*—that is to say, we like people who are *like us*. Perhaps our brains can just *tell* when someone is built the same.

⟪

As this chapter comes to a close, I hope you might be walking away with a newfound appreciation for certain people in your life. As philosopher George Santayana once said, "The family is one of nature's masterpieces." That saying really takes on new meaning when you consider that parents, children, and spouses engage our brains in special ways, bringing out high levels of oxytocin and triggering interbrain synchrony. As a result, they may provide systemic health benefits and increased potential for collaboration. This truly is one of nature's masterpieces.

As rare and precious as these bonds are, they can be jeopardized. Arguments with parents, children, and spouses can burn hot, sometimes leading to the needless collapse of these vital bridges. In these times, it's important to ask yourself what is more valuable: the content of this argument or the rare and irreplaceable bond you share?

For the good of your brain and body, take advantage of these special relationships and what they can offer you both emotionally and biologically. If you have the chance, hold these people close and cherish them. Never forget that we humans *need one another*, and there's hardly anything as impactful as those we hold nearest.

## Key Takeaways

1. Love is characterized by oxytocin release in the brain. By making interaction rewarding, oxytocin serves as a sort of neural glue that binds partners together.

2. Married people are less likely to die by cancer. This may be because oxytocin has health-promoting properties like being anti-inflammatory and neuroprotective, and even supporting wound healing.

3. Social touch can drive oxytocin release, regulate cortisol levels, and mitigate pain, depression, and anxiety. This involves the OFC and pSTS. However, the *social* nature of touch is critical. Robot masseuses don't

drive as much oxytocin release. More oxytocin is released with masseuses we *want* to bond with, like a romantic partner or a sexual interest.

4. Oxytocin is highly prevalent in parent/child relationships. This oxytocin helps form an unbreakable bond and may even protect infants through birth.

5. Two people's brains may sync up when they work together or share an experience. This interbrain synchrony occurs the most in romantic partners and parent/child duos.

To view the references cited in this chapter, please visit benrein .com/book.

# 8

BECOMING A
BETTER INTERACTOR

The Science of Likability

In the late 1990s, actress Jennifer Aniston took much of the world by storm. She was little known for much of her career—playing minor roles in films you've probably never heard of—until she was cast as Rachel Green for the hit TV show *Friends* in 1994. Suddenly her career took off. She was catapulted into incredible fame and favor. Women copied her clothing style, and a haircut even emerged called the Rachel, named after her character. Hell, neuroscientists were even using her photos when they discovered that viewing multiple pictures of the same person will activate the exact same neuron in the brain! This gave rise to the term the *Jennifer Aniston neuron*. I'm not kidding.

In retrospect, it was inevitable that Jennifer Aniston would be so treasured. The character she played had all the pieces that comprise the puzzle of *likability*. Rachel Green was beautiful, funny, and outgoing. Her lines were witty, but she was down-to-earth and relatable. As bubbly and charming as she was, she was also honest and authentic. In other words, science could have predicted Jennifer Aniston's ascent to popularity. The character

Rachel Green was scripted to be adored, and Aniston played her masterfully.

## THE NEUROSCIENCE OF LIKABILITY

In 2017, French scientist Dr. Benoît Bardy had unsuspecting research subjects pair up with a stranger for a short conversation. Little did these people know, they were actually interacting with an actress hired to play a *likable* or an *unlikable* character. In the likable condition, the pairs spoke about a neutral topic (like their hobbies and studies), and the actress was friendly and outgoing. As the paper describes, "she engaged herself completely in the conversation, listening attentively and responding properly to the participant. Her phone was switched off to avoid incoming calls or messages." She sounds lovely.

The unlikable condition looked very different. Instead of discussing a neutral subject, the pairs spoke about controversial topics like gay marriage. Once the actress had gathered the subject's opinion on the matter, she purposely voiced the opposite stance. She avoided eye contact and was inattentive when the subject spoke. She even set an alarm to make her phone ring during the interaction, and then after it went off, she continued texting.

Let's be honest: This person kinda stinks. I'm not afraid to admit, I would find these behaviors pretty off-putting, and I wouldn't particularly *like* her . . . and that's exactly the point. She was purposely doing everything *wrong*, being disagreeable, inattentive, and rude. Meanwhile, in the likable condition she was friendly, engaged, and polite. I, for one, would *much* prefer the company of the *likable* character.

Just as Jennifer Aniston's character was scripted for success, so was the likable actress. The opinions we form of one another are not totally random; they're guided by certain principles that science has characterized fairly well, and Dr. Bardy's team nailed it.

This process of forming impressions of others involves brain areas like the prefrontal cortex (PFC). As you may recall from chapter 2, this is the region that Sandi, the worker at the garden store, likely used when she formed her impression of me. Within the prefrontal cortex, there are two smaller subregions that play special roles in this process: the ventromedial[+] prefrontal cortex (vmPFC) and the dorsomedial prefrontal cortex (dmPFC). The vmPFC comes online when we look at attractive faces, and shows greater activity when people rate others as more likable. It seems to be an important hub where your brain computes how favorably you view someone. The dmPFC—just an inch or two closer to the top of your head—seems to play a similar role, as its activity tracks how charmed we are by others. In one fascinating study, single people sat in a brain scanner while they looked at pictures of other singles. They were essentially having their brain activity measured while swiping in a dating app. In an exciting twist, the singles got to speed-date each other after viewing each other's photos. Amazingly, it was found that the activity in someone's dmPFC could predict their later dating decisions. If their dmPFC lit up when they saw that person's picture, they were more likely to pursue that person for a second date. If singles could only sit in an fMRI machine while they swiped on Tinder, the scans might predict the outcome of their dates!

The dmPFC isn't just infatuated with appearances, though. It's also involved in more sophisticated tasks like judging someone's characteristics. If I had snapped my fingers in the garden store and scrambled the activity of Sandi's dmPFC, she might've had a harder time judging what traits I have, like whether I'm selfish, outgoing, or trustworthy.

When someone meets you, their brain takes in tons of information about you and churns it all together through the PFC to spit out a judgment. *Do they like you or not?* This question is one that has tormented humans globally for millennia. *Will they want a second date? Did I make a good impression on the parents? Will I get the job?*

Humans are obsessed with our reputations, and that's understandable. It's simply in our genes. We have nightmares about getting our teeth knocked out, being naked in public, or failing a test—all things that pose a threat to our reputation. Why? Because as a social species that once relied upon one another for survival, being likable was crucial for staying alive. If you were disliked enough to be kicked out of your tribe, you were as good as dead. A human alone was easy prey in a prehistoric world.

It's too late for our unlikable ancient ancestors, but there's good news for you. Researchers have been studying the *psychology* of likability for decades, and we have a good grip on the factors that influence how likable someone is. Certain traits seem to tickle the PFC and lead to more favorable judgments, while others do the opposite. I think it's valuable to learn about these factors for three reasons. First, you must have successful social interactions if you hope to reap the benefits they hold. Being well liked allows you to deepen your relationships, which may provide secondary benefits for your health and well-being.

Second, it's generally productive for relationship building if people understand how their behaviors affect others' impressions of them. That's not to say we should be *hacking* and *optimizing* our relationships, but it would be better if we didn't do annoying stuff all the time that might rub people the wrong way. As you recall from **Hard Truth No. 3, there are many internal factors in the brain that divide us.** We would each be well served to understand the factors that drive wedges between us and make us unlikable. This book is all about making the most of our interactions, and making a good impression on others is a significant piece of that.

Third, research shows that well-liked people enjoy tons of surprising benefits. For instance, when students like their instructors, they rate them as better teachers and are more likely to comply with their requests. In mock criminal trials, testimonials from less likable experts are deemed less persuasive. In the Chinese live-streaming market, streamers rated as more likable received more tips and had higher levels of viewer participation. In many domains of life, it helps to be likable. So why not work on ourselves a bit to deepen our connections and attract more success?

Plus, being disliked does you no good. You might recall from chapter 4 that people feel less empathy for those they dislike, and men even show pleasure-related brain responses when watching them receive painful electrical shocks. Yikes. I guess it stinks to be an asshole (no pun intended).

By this point in the book, I sincerely hope you're feeling inspired to engage in social interactions for the sake of your well-being. I'm excited for you to implement those changes, and when those interactions *do* happen, you may feel more confident if you're prepared with an understanding of likability. Let's work through the key variables that feed into likability, and we're going to do so in the only way that feels appropriate: by putting ourselves in a lifelike situation.

## UNDERSTANDING THE FACTORS
## THAT GUIDE LIKABILITY

Imagine that you've recently entered a new romantic relationship, and things are getting pretty serious. They're a great partner for you, and you've never been happier. It's a sunny, crisp Saturday afternoon in fall, and you're headed to a local coffee shop together. Today is a big day, one you've been thinking about all week. You're about to meet your partner's best friend, Larry. You've heard a lot about Larry—how close he is to your partner, how they went to school together, and all the rest. Naturally, you want to make a good first impression. As you walk into the coffee shop, you can't help your nerves. You really don't want to screw this up.

As you sit down at a table, your partner looks at you with an anxious expression. Suddenly you realize they're hiding something. "Hey . . . I hope you won't be mad," they say, "but . . . I haven't told Larry about you yet. He has no idea I've been seeing someone. I wanted it to be a surprise."

You open your mouth to ask if they're serious, but it's too late. Over your partner's shoulder, you see the door swing open as Larry walks in. With a deep breath, you steady your nerves and get ready to put your best self forward. Here we go . . .

## The Uncontrollable

There's good news and bad news in this situation. The good news is, you have control over many of the things that influence your likability. Phew! The bad news is, some of these things are out of your control. Speaking of which, here he comes. Shit. Just be natural!

"Hey," Larry says, approaching your partner. He gestures toward you with a soft smile. "Who's this?"

You feel a jump of nerves as you awkwardly outstretch a hand to introduce yourself. "Hi, Larry, it's so nice to meet you. I've heard so much about you. My name is . . ."

Wait. What if he judges you based on *your name*? Is that something that matters? Surely that can't affect his perception of you, can it?

Actually, it may not be out of the question. In a 1995 study, subjects were asked to choose which positive labels (like "responsible" or "tolerant") they associated with a list of names. Half of the names were common (like Michael and Jennifer), while the rest were relatively uncommon (like Rodney and Joanne). Curiously, the students associated more than twice as many positive traits with common names than with uncommon names, suggesting that rarer names may be viewed less favorably. Consistent with this, other studies show that elementary students with common names are more likely to be popular and receive higher scores on essays. Meanwhile, people with *harder-to-pronounce* names are judged as less trustworthy and more dangerous. Considering that it's the first thing Larry will learn about you, it's plausible that your name could contribute in some way to his impression.

Oddly, your name may also influence how *attractive* Larry thinks you are. When headshots are paired with names that are rated as more attractive (like Danielle or Alexander), the model is judged to be better looking. Meanwhile, when photos were accompanied by unattractive names like Tracey or Kenneth (sorry, Traceys and Kenneths—that was according to the study, not my opinion!), they were judged as less attractive. This effect was particularly strong for women. On average, women were rated almost 10 percent more attractive when associated with a favorable name.

So as you extend your hand and introduce yourself to Larry, should you use a fake name? Probably not—but that's just my opinion. I don't think life gets any better for people who sculpt themselves into something they're not for the sake of pleasing others. Plus, science shows that there's a positive relationship between *authenticity* and likability: People are better liked when they are perceived to be authentic. In other words, be yourself. Sacrificing your true identity for the sake of likability may be a fruitless endeavor that only bites you in the ass.

As you look at Larry and tell him your (real) name, you can't help but notice that he's a handsome guy. Soon you find yourself thinking about your own appearance and wondering if it matters.

It sure as hell does. Being good-looking is linked to a shockingly long list of perks. Attractive people are judged as more likable, more trustworthy, more sincere and altruistic, and even more intelligent. This is sometimes called the *halo effect* or the *beautiful-is-good stereotype*, and it bleeds quite deeply into the way we judge and treat others. One study found that more attractive workers were more likely to get promotions and raises. Meanwhile, in college classes taught by male professors, better-looking female students got higher grades when classes met in person versus online. Pretty shocking, right? Between your name and your appearance, much of your impression on others is predetermined before you have any say in the matter.

Now, I should acknowledge that your name and attractiveness technically *aren't* uncontrollable; there are ways to change your name or alter your appearance through aesthetic operations. I don't fault anyone for doing those things, but I ask that you please not do them on account of this chapter. The point here isn't to chisel yourself into the most charming, magnetic human possible through any means. It's about understanding human nature and psychology. Plus, these modifications may compromise your authenticity, and we already know how that affects likability.

With that said, there are many other traits that we *can* adjust to improve our impression on others without sacrificing authenticity. I believe that if someone can recognize their own best (or worst) qualities, they might learn to feel more confident in interactions by spotlighting (or moderating) those traits. In an ideal world, we can harness this knowledge to be stronger interactors while remaining true to ourselves.

## The Controllable

Now that you are a few minutes into your conversation with Larry, things are starting to feel more comfortable. You've managed to settle into your chair and immerse yourself in the conversation, largely forgetting about your nerves. It crosses your awareness that you're sitting kind of weirdly—your arms are dangling off the sides of your chair limply. Then suddenly it all slams back into you. *What the HELL am I doing with my arms?*

Body language can be a point of discomfort for some, and science suggests it can indeed make a difference. Although we may not notice it, most interactions involve a lot of movement. You might waggle your arms around while telling a story, tilt your head back in laughter, or roll your eyes in disbelief, all while alternating between sitting cross-legged or cross-armed, touching your face, and so on. But what's not often recognized is how much of this movement is *copycatting*. Next time you have the privilege of observing two people in conversation, watch their bodies. They'll often mimic each other's posture and adopt similar positions. You can even test this out yourself. In the middle of a conversation, try folding your arms or placing a hand on your hip. Oftentimes your partner will soon find themselves doing the same. Although this mirroring may seem intentional, it usually happens beneath our conscious awareness and seems to be a social tool for *building affiliation*. Studies show that people mimic others more after they've been excluded from the group, presumably to get back into good standing with their tribe. What's more, it works. Subjects rate others more favorably if that person has mimicked them, of course without being consciously aware that they were mimicked. When researchers had an actor mimic the posture and movements of subjects while they worked together on a task, the subjects reported liking the actor more. So don't fret if you notice yourself mimicking Larry; it's your brain's sly attempt to gain his favor. Also, if you notice him mirroring *your* movements, it could be a good sign.

On the other end of this spectrum, a lack of mimicry can be detrimental. People with high levels of social anxiety tend to spend more time monitoring their *own* behavior, in fear of making a mistake or leaving a bad impression. As a result, they focus less attention outward on their partner and thus engage in less mimicry. This has been linked to lower ratings of likability, suggesting that high mimicry doesn't just build rapport—low mimicry can also harm it.

Now, to deliberately put this into practice would be somewhat ridiculous. If you mirror Larry's every move, he will probably notice and think it's weird. If you want to affiliate with him, your body will probably do this on its own. Don't overthink it.

Just as your partner is wrapping up a story, you feel Larry gently tap your forearm. "So where are you from?"

For a moment, you're caught on the sensation of this unexpected touch, which was surprisingly endearing. Although it may sound odd, making physical contact can drive social connection in a powerful way. Social touch is generally a positive, friendly thing to do—just think of how it feels when a friend grabs your arm while laughing. Studies show that female servers receive larger tips from customers if they touch them while serving them, and people are more likely to help a stranger pick up dropped materials if that person had touched them in conversation, even just for one second. Considering what we learned in the last chapter—that touch drives oxytocin release—it's no surprise that touching someone can help build bonds and affiliation. Don't force it, but if you see an opportunity to reciprocate Larry's touch in a friendly and noninvasive way, it may help your case.

Snapping back into the present moment, you realize that Larry's been saying something for a while, and you suddenly become *painfully* aware that you've been holding extended eye contact with him. It feels a bit intimidating, so you avert your eyes by looking at your partner. Then you turn back to Larry and lock eyes again. A few seconds pass. You wince with uncertainty. Is this okay?

Recall from the beginning of this chapter that Dr. Bardy's team instructed their unlikable character to avoid eye contact. That's because eye contact is indeed good for connection. After holding eye contact, people report higher respect, affection, and liking for each other. You may also recall from the last chapter that interbrain synchrony peaks during eye contact. Looking into someone's eyes gives you a window into their internal state, and that vulnerability is quite supportive for connection.

If you find eye contact uncomfortable, there may be a simple cheat code here: Just look at the person's mouth. It turns out, people will perceive the same amount of eye contact regardless of whether you're focusing on their mouth or their eyes (but you do have to be at least three feet apart). I just tested this trick with my wife, and it really does work. She had no idea. This eye contact illusion is an easy one to try, and don't worry—you won't look peculiar. Even if they do notice, it's very normal to look at someone's mouth in conversation. Surely you've done this to better hear someone in a noisy bar or party. Ideally, it's probably best to alternate between looking at Larry's eyes and mouth.

Larry's impression of you can also be influenced by your facial expressions, and when you're in doubt, smiling is a safe bet. Smiling faces are judged to be kinder and more honest, friendlier, politer, and even younger. Similarly, people are judged as more likable when seen laughing. Smiling people are also judged to be more affiliative—that is, more interested in building relationships. This could be why people awkwardly smile at each other when passing in a hallway. There's nothing to be smiling about, unless you deeply enjoy walking through hallways. Rather, this smile may be a primitive way of acknowledging the other person and wordlessly signaling to them, "I'm friendly, and I'm not going to attack you."

In contrast, people making expressions of anger or disgust are viewed as much less affiliative. This could explain why people afflicted with "resting bitch face" are sometimes judged to be unfriendly or cold. The human brain may have a bad habit of generalizing someone's facial expression to their entire personality.

These tips might be able to help you form good impressions on others, but of course they are best done in moderation. If you mimic Larry's every move, touch him constantly, never stop smiling, and refuse to break eye contact, he's probably going to be completely terrified of you. Don't overcorrect your behavior to fit these guidelines—rather, use it as an opportunity to reflect on what you could work on in your interactions.

⬯⬯

Body language is clearly a vital piece of this puzzle, but we also can't forget about *language* language. Conversation is the backbone of connection, and it certainly matters how much you talk, what you say, and how you say it.

I have a very specific question for you. If you want to make a good impression on Larry, how much do you think you should talk? Be precise: What percentage of the interaction should *you* be speaking?

Most people will guess somewhere around 40 percent. Did you also guess something less than half? I find it so interesting that people think they should speak *less* to be liked *more*. In truth, research suggests the opposite: It's better to *speak more*. One study paired up college students for a short conversation and assigned each of them an amount of talking to do. Some pairs got an even 50/50 split, while others were as lopsided as 70/30. After the conversation, everyone judged how much they liked their partner. In the end, those who spoke more (50, 60, or 70 percent) were rated higher than those who spoke less (30 or 40 percent)—exactly the opposite of what everyone expected. Of course, there's an upper limit to this. Don't speak so much that your partner feels muted. Be sure to leave plenty of room for them.

Most people know that everyone likes to talk about themselves, which might be why we assume we should speak less and leave more room for the other person to fill. However, all good conversations are a volley, and there's a lot of joy in keeping the ball moving back and forth. If you feel the need to pipe down, a good alternative is to ask questions. While this keeps the ball moving, it may also improve your impression. When singles in speed-dating conversations asked more follow-up questions, they were more likely to land a second date. Great conversations aren't all questions, though; overdoing it can quickly make things feel like an awkward interrogation. It's best to find a natural balance between asking, listening, commenting, and letting things flow organically.

As Larry is talking, you notice that he's tapping his foot—a habit that you also do. For some reason, you find this charming. That's because people tend to like others similar to them (remember *homophily*?). It could be a similar attitude, a similar personality trait, or even just a shared name or birthday, but if it draws you together, you're prone to liking them better. For instance, speed-dating couples are more likely to request a second date when they match each other's speaking style. This reminds me of self-other overlap and the brain's empathy disposition. When you view someone as *like you*, your brain seems to give them all sorts of benefits.

The influence of subtle similarities can even extend to shared experiences. For instance, one study found that when people watched a movie together, their bodies tended to synchronize in certain ways. They would smile at the same times in the show, and get sweaty and nervous simultaneously. More important, the pairs who showed more similar physical responses to the show *felt more connected afterward.* These synchronized physical responses could reflect homophily on a deeper level, revealing a shared alignment of emotional and physiological responses. Thus, watching TV could be a meaningful way to bond, even though it doesn't involve any direct interaction. If you're smiling and you look over and they are, too, it's probably a good sign.

◯◯

Back at the coffee shop, things are going well. You're really warming up to Larry, and you can see why your partner is so close with him. He's a cool dude. But then something terrible happens. Larry pulls out his phone and starts texting.

As we learned in chapter 6, people enjoy interactions less and feel less socially connected when their partner uses a phone (also known as *phubbing*). This is exactly what the unlikable character did in Dr. Bardy's experiments: She got a loud notification, then became absorbed in her device. We now know that this is a bad thing to do, but there is another issue at play here: You may be blind to your own phubbing. People tend to believe that when *they* use their phone, it's for a good reason. They overestimate their ability to multitask and consequently excuse their own phubbing. However, they still perceive it negatively when others do it. Call me old-school, but I'm a big fan of the "phones away at the dinner table" rule. Let's get back to connecting and forget about our devices for a few minutes. They can wait.

As you're gathering the courage to call out Larry's rudeness, he looks up with a smile and puts away his phone, abruptly blasting into an engaging and funny story. You can't help but notice how utterly *expressive* he is, and you find him winning you over again. Indeed, studies show that more emotionally expressive people are better liked. These are folks who rate themselves higher in questions like "I often touch friends during conversations" or "I can easily express emotions over the telephone." Considering how much information is carried through these social cues and expressions of emotion, maybe these people are better liked because they're easier to read. Putting your emotions on the table makes it less work to understand you, and that can be refreshing.

Similarly, research shows that more extroverted people are better liked, but with a caveat: They must also have *high emotional intelligence*. An outgoing, energetic person can be a joy to talk to, but that delight can be wiped out if that same person is overbearing. Having high emotional intelligence allows people to track the emotions of others and recognize the impact of their own behavior. That's valuable for ensuring that extroverts apply their social energy in a way that's tasteful and considerate. Larry is lucky in that he seems to have both . . . But he better keep that damn phone in his pocket.

At last your partner signals that it's time to go, and you cheerfully say your goodbyes. You think it went well overall, but you can't help feeling unsure. On the way home with your partner, you let your insecurities get the best of you. "Do you think Larry liked me?" Your partner reassures you by saying that it truly couldn't have gone better, but for some reason you're not buying it. You just can't shake this feeling that he didn't like you as much as your partner insists.

Monitoring your own likability is a difficult task. It turns out, people are very bad at guessing how much people like them. We're now a long way from the five *bad predictions* I laid out in chapter 3, but I suppose this would be **Bad Prediction No. 6: We underestimate how much others like us.**

This phenomenon is called the *liking gap*, named by psychologist Dr. Erica J. Boothby. In Dr. Boothby's studies, people consistently underestimated how much others liked them after a conversation, and assumed the other person didn't enjoy the conversation as much as they actually did. The moral of this story is: Don't be so hard on yourself. Larry—and others—probably like you and enjoy your company more than you think. To spend your energy and time vying for approval is a waste of resources. Hell, it might even prevent you from reaping the benefits of these social interactions. In the end, it's best if you simply be yourself.

Seriously. You're probably doing a lot better than you think.

### Key Takeaways

1. Making judgments of others involves multiple regions of the prefrontal cortex. Both the vmPFC and dmPFC are more active when we like someone.

2. Likability matters. It's important for building social connections, and well-liked people reap many real-world benefits.

3. Some factors that influence likability are *uncontrollable*, like one's name and appearance.

4. Others are *controllable*, such as body language, eye contact, conversation style, facial expression, touch, cell phone use, and others.

5. People tend to underestimate how much others like them. This is called the *liking gap*.

To view the references cited in this chapter, please visit benrein .com/book.

———— ○ ————

# 9

# UNDER THE INFLUENCE

How Drugs Manipulate the Social Brain

One of the few downsides of studying neuroscience is that it strips away much of the magic and mystery of being alive. It reveals to you that almost everything—from déjà vu to drug-induced visions and hallucinations—can be explained by chemical events unfolding within the brain.

At the end of the day, everything you've ever experienced has been a pattern of electrochemical signals in your brain. When you pet your childhood dog for the first time, swung at a ball in gym class and missed, or learned that you were promoted at work, there was a distinct pattern of activity emerging invisibly beneath your skull. Each thought, emotion, belief, sensation, opinion, or reaction is accompanied by a specific neural arrangement. If we could capture that pattern and bottle it, it could serve as a souvenir representing your conscious experience in that moment. I'm unsure whether this is a sad truth or a beautiful one, but it *is* the truth: **Everything you experience boils down to a pattern of chemicals in your brain.**

To back up this claim, let's look at the five senses. Information from the outer world enters your body in many forms: *light* shining on your retinas, *sounds* vibrating into your ears, *textures* dragging across your skin, *odors* drifting into your nostrils, *flavors* spreading across your tongue. Upon meeting your body, these *physical* signals are swiftly translated into the brain's native language: *chemical* signals. Noises, for instance, are vibrations that oscillate at a certain frequency. When these physical sound waves smack into your ear canal, they cause some biological machinery in your inner ear to vibrate. Those vibrating structures are pressed against a fluid-filled space called the *cochlea*, and their vibration creates waves in that fluid—similar to how the water in a pool ripples when you slap your hand on the surface. Those waves rush by some tall structures called *stereocilia*, which causes them to bend, like a strong tide pushing seaweed to one side. When those stereocilia bend, they allow *electrically charged ions* to flow into the cells they're attached to, which triggers the cell to release neurotransmitters onto the auditory nerve. In mere milliseconds, those *physical* sound waves from the outside world were translated into *chemical* signals (in the form of neurotransmitters) that the brain could interpret. Amazing, if you ask me.

As impressed as I am by the brain, I also feel a bit sorry for it. It would be *isolating* to be a brain, wouldn't it? The poor thing is locked away in total darkness in the head, using a chemical language that's completely foreign to the world beyond. In the outer world, a drop of dopamine or a dollop of oxytocin means nothing. But in the brain, these chemicals tell a rich story. They're all the brain knows.

However, it turns out that many naturally occurring substances *do* speak the brain's language and, as a result, can manipulate its internal communication systems. For instance, psilocin, the active compound in psychedelic mushrooms, behaves like serotonin in the brain, binding to certain serotonin receptors and activating them. Meanwhile, cocaine—derived from coca leaves—hijacks dopamine transporters, causing neurons to pass dopamine around in excessive volumes. Drugs like these, which figuratively "speak" the brain's language, toy with our neurochemistry in ways that can alter perception. Like an unruly guest running rampant through the rooms of your home, these drugs weave through your brain, flipping light switches on and off and tinkering with your neural activity. If you had taken psychedelic mushrooms before petting your childhood dog, its fur may have felt different on your fingertips. If you were on cocaine when the announcement was made that you were promoted at work, you may have felt particularly elated and perhaps celebrated a bit too expressively. By whispering messages to your brain's chemical systems, certain drugs can profoundly alter your experiences and behaviors.

Social interactions are certainly not immune to these effects. Just as sound waves are translated into chemical signals, so are your social exchanges. Everyone you know and your conversations with them also imprint a bespoke chemical signature within your brain. This means that certain drugs can just as easily influence our *social behaviors*, flipping switches in brain areas that process social information and underlie connection.

It may feel unexpected to have a chapter about drugs in a book about social connection. But any neuroscientist who studies these compounds will tell you that it's actually quite appropriate. Many potions from the outer world *do* act directly on the social brain, coloring our interactions in ways that can be helpful or unhelpful. I believe that drugs are one of the *external* factors that don't get enough attention in the mainstream conversation about social interaction. We know that social media and political polarization are driving us apart, but what about the substances we ingest nearly every day? While some drugs may create *distance* in our social interactions, others may *enhance* social connection. Understanding which do which (and why) may be of incredible value. However, please do not use this chapter as a guide for using these drugs. Always speak with your medical provider before making any changes to your prescriptions, as some drugs like SSRIs can have severe withdrawal effects if not tapered properly. Your neurochemistry is delicate and must be handled carefully. Please treat it with care and caution.

## TWO BRAINS WALK INTO A BAR

In a subtle, unspoken way, alcohol has grown into the official drug of social interaction. So much of modern culture is built around its use as a social lubricant, an elixir that cures us of awkward, clumsy exchanges. It's like oil to the engine of the social brain, removing much of the friction two people may typically encounter. Groups pregame events to ensure everyone shows up with a buzz. Couples clink together glasses as they ease into the complexities of dating. At last call, singles order one more and scan the room eagerly. Under alcohol's influence, people connect in ways that are unrestrained and sometimes reckless. We share more about ourselves, laugh more, engage in more conflict, and become more sexually promiscuous, among other things. There is a common tie that links each of these behaviors: They're what we might call *uninhibited*.

The average sober person has a great deal of top-down control over their behavior. Their brain's sophisticated thinking centers can override their more primal urges. They might feel an impulse to share a deeply personal story, draw the attention of others with a loud laugh, fight the guy who bumped into them, or sleep with a stranger, but before they do, a convincing voice warns them to pause and reconsider. Sobriety comes with the ability to suppress bad ideas before they become bad decisions. But when alcohol is on board, that restraining voice seems to be subdued. Those top-down commands shrink to top-down whispers. Why does alcohol do this to the brain?

You've probably heard that alcohol is a nervous system *depressant*. That doesn't mean it makes people depressed—it means it lowers activity in the brain. If you've read the appendix entry on brain signaling, you'll recall that neurons can send one another *in-*

*hibitory* signals, which bring down one another's activity. The most common inhibitory neurotransmitter in the brain is GABA. When GABA receptors are activated, they open like tiny pores on your neurons and allow negatively charged ions to flow in. This lowers the electrical charge of that neuron, reducing the likelihood that the neuron will fire. This is how GABA suppresses brain activity.

Alcohol is a depressant because it tickles GABA receptors in a way that makes them let in more negative ions. As a result, your brain's natural GABA signals become stronger and *more* inhibitory than usual. In other words, alcohol makes your neurons better at shutting one another up. In turn, that can make you worse at something you're usually quite good at: *thinking*. When neurons in certain brain areas are silenced, high-level cognitive processes can become less accessible, and your behaviors may turn reckless. With your brain's computing power diminished, the thoughtful voice in your head grows muffled, and your actions become more reliant on impulses and urges.

Some anxiety medications like Xanax or Ativan work the same way. They enhance GABA signaling in *fear*-associated brain areas like the amygdala, which can soothe one's nerves by softening the intensity of their emotional reactions. In similar fashion, alcohol has been shown to turn down amygdala activity. When tipsy people look at angry or fearful facial expressions, they show weaker responses in the amygdala. This means that alcohol may *weaken* our emotional responses to negative social signals, which makes perfect sense to me. When you lean in for the kiss and see that you're being fully rejected, it may not sting so bad. You may not feel as remorseful when your tasteless joke is met with faces of disapproval. Processing negative social signals is important for keeping our behaviors in line, but with those regulatory guidelines thrown in the trash, we're much more likely to make dumb or inconsiderate decisions. So next time you've had a couple of drinks, try to keep this in mind (if your hampered neurons will let you).

Perhaps because of this overlapping mechanism between alcohol and anxiety meds, drinking tends to take the edge off in social settings. People feel less anxious on blind dates after having a drink, and one small study found that social anxiety levels dropped by 4 percent for each drink a research subject consumed. This anxiety-blunting effect might also be the reason why people turn to a stiff drink after a stressful day.

Together, these two findings can explain a lot of the silly human behavior we see in bars. When your brain's reactions to negative social signals are muffled, you're prone to all sorts of social miscalculations. Stack reduced social anxiety on top of that, and there's no question that your decision trees will look different. Should I share my extremely personal problems? *Sure, who cares!* Is it a bad idea to hook up with my coworker? *Nope, how could that possibly backfire?!* Would it be rude to interrupt this boring story? *If they get upset, I won't feel a thing!* Should I invite this stranger back to my house? *I can't think of a single reason why not!*

Although it might feel like we become *better* at socializing after a few drinks, I would actually argue that it makes us less effective interactors in several ways. We become emotionally detached from the interaction, reacting less to our partner's social cues and experiencing muted anxieties. While that might be helpful in certain contexts—like taking on a difficult but important conversation—it can also create unhelpful emotional distance in our interactions. As hampering as social anxieties can feel, it's advantageous to be able to recognize and respond to others' social cues, especially when those cues are telling us what *not* to do. When we disregard social consequences, we boost the likelihood that we will embarrass ourselves or insult someone through our actions. I believe that social interactions are best when both parties are cognitively engaged, thoughtful, and considerate of each other's emotions. Alcohol takes away many of these valuable features and at high doses may also reduce emotional empathy, which could lead to even more insensitive actions.

Alcohol's negative influence on one's social life may worsen if drinking turns into a long-term habit. For instance, alcohol use disorder has been linked to challenges evaluating facial expressions and vocal tone. Suppose it's a cold day and your friend is wearing a wool hat and a big puffy scarf, leaving *only* their eyes visible. Based on their eyes alone, do you think you'd be able to judge what they were thinking or feeling? This is called Reading the Mind in the Eyes, and research shows that people with alcohol dependence struggle to pull this off. They also show reduced activity in empathy-related brain areas when attempting to do this, suggesting that chronic alcohol use may impair the brain's empathy systems. To make matters worse, people with alcohol use disorder also show less of that interbrain synchrony.

Considering that alcohol is (unofficially) the official drug of social interaction, this is pertinent information. A huge segment of the world's population drinks—often to *enhance* their social interactions—but how many are aware of these social *consequences*? Allowing a rampant guest to flick off the lights could have real consequences—some good and some bad. We would be well served to keep close tabs.

## HOW PAINKILLERS MAY BE SECRETLY IMPACTING YOUR SOCIAL LIFE

I'm sure you didn't expect to see this here. What do painkillers have to do with socializing?

Surprisingly, actually quite a bit.

Acetaminophen is a cheap, effective, and widely accessible painkiller taken by millions of people every day. Most people know it's a quick and easy way to soothe a headache or sore muscles, but not everyone is aware that it can affect how the brain processes social information. That's because it doesn't just dull *physical* pain: Research shows that it can also moderate *social pain*.

At UCLA, researcher Dr. Naomi Eisenberger (who we also ran into in chapter 6) ran a clever study where college students were instructed to take a pill every day for three weeks. Some students got acetaminophen, while others got only a placebo—but of course nobody knew what they got. Over those three weeks, the students kept a record of how much social pain they'd experienced each day. Shockingly, the students who got the painkillers reported having their feelings hurt less.

To make sure this wasn't just a coincidence, Dr. Eisenberger's lab turned to brain imaging. A new set of students took acetaminophen every morning and evening for three weeks, before coming to the lab to play Cyberball—that computer game of virtual catch. As the ball was being passed around, this time *they* were excluded. This was intended to induce social pain, as the researchers wanted to see if the acetaminophen would alter how the *brain* responded— and it did. The students who took painkillers showed *less activity* in the dorsal anterior cingulate cortex and the anterior insula, both of which are involved in processing social pain. They also showed weaker activation of the amygdala, potentially suggesting a more regulated anxiety response. Incredibly, it seems that acetaminophen can dull social pain by hushing down the brain areas that process it.

You may also recall from chapter 4 that painkillers can reduce how much *empathy* people feel for others' pain. Subjects who have unknowingly taken acetaminophen find it less disturbing to read about painful experiences and also rate the character's pain as less severe. The same is true for socially painful events, meaning that painkillers reduce empathy for social pain as well.

How is this even possible? How can painkillers block empathy for the physical and social pain of others? The answer is surprisingly simple. To feel empathy for someone's pain, you must engage the brain areas that encode the emotional part of pain (like the anterior cingulate cortex and insula). The painkillers may make it harder to bring online these pain-processing regions, thereby blocking your ability to understand the other person's experience. Indeed, Dr. Eisenberger's earlier study showed that acetaminophen reduces activity in both regions.

What might surprise you, though, is that acetaminophen doesn't just blunt empathy for pain—it also reduces empathy for *positive* experiences. When people read positive, uplifting scenarios (such as stories about a couple getting engaged or a worker getting a raise) after taking the drug, they experience less personal pleasure and weaker empathic feelings.

If you're afraid that all the acetaminophen you've taken over the years has permanently blunted your empathy, don't worry—there's no evidence for that. People who take painkillers more often don't show any differences in empathy. Unlike what we saw in chronic alcohol use, it seems that painkillers influence empathy *only* when they're *actively* in your system, but don't create long-lasting effects. This makes sense through the explanation that painkillers block empathy by temporarily suppressing the brain's pain systems. Just as your pain comes back when the painkiller wears off, empathy may also be restored once those brain areas regain their typical function. However, this may not be the case for all pain relief drugs; long-term opioid use has been associated with reduced emotional empathy, even when the drug is not in a person's system. It's unclear exactly what is happening there, but the point is that not all painkillers have equivalent effects. Opioids also act through a different mechanism in the brain than acetaminophen, which might explain these differences.

I share this information about acetaminophen for good reason: It clearly has serious implications for real life. If painkillers can reduce social pain and blunt empathy, they may be silently slashing holes in the complex webs that make up our social lives. For people with conditions like chronic pain that require frequent use of painkillers, there may be off-target effects in the social domains of life. Being aware of these potential consequences may enable you to safeguard against them. You can't solve a problem that you don't know exists.

Alternatively, the social effects of painkillers may not be as hidden as I propose. When I've spoken about this on social media, many people have shared that they've tactfully used painkillers for this exact reason: to ease the soreness of social pain in tender times, like during a breakup. Medical professionals have admitted to taking acetaminophen before shifts to reduce their empathy for suffering patients, thereby avoiding discomfort themselves. These cases involving deliberate use of painkillers are interesting, but there are plenty of other situations where this emotional blunting would be *unhelpful*.

Remember: Empathy is valuable because taking on someone else's pain motivates you to step in and help them. This is one reason why humans survive better in groups: because we help one another. Our sensitivity to one another's emotions is partially responsible for our success as a species.

However, we may be losing some of this when we take painkillers. When you see someone being bullied, you probably won't care as much with acetaminophen in your system because *you* have trouble relating to their social pain and thus feel less distressed as an observer. You might be less likely to pull over and help someone in a broken-down car if you're cruising past on painkillers. If everyone took acetaminophen every day, we would almost certainly inhabit a rougher, less generous world. Although it can be uncomfortable, it is an immense blessing that we *are* sensitive to the pain of others. To rid ourselves of this ability is to discard one of the tools that our ancestors relied on for survival.

## CAN ANXIOLYTICS LIMIT ALTRUISM?

Imagine you're attending a close friend's wedding and it's time for the best man to stand up and deliver his speech. He rises clumsily and stumbles over to the mic, and it becomes painfully clear that he went a little overboard at cocktail hour. This guy is plastered. Unfortunately, he plants himself directly beside you as he begins addressing the crowd. Oh boy, this is going to be a disaster.

As expected, he's a sloppy drunk mess, slurring his words and saying some *very* inappropriate things about the couple. Looking over at the newlyweds, you can tell from their expressions that they are getting uncomfortable. Then you notice a shift as the best man starts to become nervous, too. He wipes a bead of sweat from his glistening forehead as his speech stalls. As you watch this guy obliterate his reputation in front of three hundred plus people, *how does it feel?*

Thanks to empathy, you would probably feel very uneasy in this situation. If it gets bad enough, you might eventually choose to stand up and snatch the mic from him to make it stop. Let's assume you do that. Now here's the big question: *Why* would you do this?

On paper, this may look like an act of altruism: a selfless behavior that is purely for the good of others. Thanks to empathy, you could sense that his speech was making *everybody* uncomfortable, including him. By putting a stop to it, you were preserving the happy couple's special day and stopping the best man from digging his own grave. But is that really the whole truth? While you may have had those intentions in mind, it's a safe bet that you were probably also acting in *your own* interests. Watching him bomb was horribly unpleasant. If you could just stop the atrocity, it would relieve *your* distress.

Some research supports the notion that acts of kindness are sometimes done for *our own* good, born from a desire to relieve our own discomfort. For instance, recall from chapter 5 that if a rat is trapped in a cage that can be opened only from the outside, other rats will step up and free their buddy. However, when researchers gave those heroic rats a *benzodiazepine*—a class of drugs used to treat anxiety—there was a sharp reduction in how much they helped their friends. This suggests that they may have been helping primarily to relieve their *own* uneasiness. With those negative emotions blunted, they were less motivated to help their friend. To prove this was really the case, the researchers placed a bar of chocolate in the cage instead of a rat. This time, the anxiety-free rats had no problem popping open the cage. It seemed that with their anxiety blunted, they weren't as motivated to step up and release their friend—but they were still willing to work for a little sweet treat.

To be sure, this doesn't mean that altruistic acts are purely selfish. While helping others may alleviate our own suffering, that distress stems from witnessing *their* suffering. This system is thus inherently prosocial—not sefish—and is meant to encourage helping behaviors by giving us a personal stake in the matter. It's yet another evolutionary tool that makes us work better in groups. But when our ability to take on those negative emotions is blunted (courtesy of an anxiety pill), we may be less intrinsically motivated to help.

This makes me wonder if you would be less likely to stop the best man's catastrophic speech if you were on an anxiety medication. What if you were throwing back shots at the cocktail hour, which we learned can have similar effects on anxiety? Or what if you had taken some acetaminophen before the wedding, making you less sensitive to his social pain? It is starting to become quite clear how various drugs can influence our social perceptions, invisibly shaping our daily exchanges.

## EMPATHY AND ECSTASY

So far, we've taken a tour of three drugs that *blunt* social functions: alcohol, acetaminophen, and benzodiazepines. To me, it's easy to see how a drug could *disrupt* the social brain. In fact, that seems much more conceivable than a drug *enhancing* brain function. It's an unfathomably complex organ, and you would think everything is already operating at peak performance. In comparison, just think about how much easier it would be to screw up your iPhone by tinkering with the circuit board than it would be to make the device work *better* or *faster*. Is it even possible that a drug could push the brain beyond its usual efficiency? To reach new heights seems like a tremendous challenge.

However, some substances *have* been shown to enhance social experience . . . and MDMA is one of them.

MDMA (whose full name is actually 3,4-methylenedioxymethamphetamine—say that five times fast) is a man-made drug that also goes by names like molly or ecstasy. With its modern reputation for being used illegally in party settings, you might guess that MDMA's origin story looks like something from *Breaking Bad*: A guy in a van cooks something up that makes him feel amazing, and he starts selling it at nightclubs. But no—in truth, MDMA was first synthesized by none other than Merck Pharmaceuticals in 1912.

Merck had no idea they'd created one of the most intensely psychoactive compounds on the planet, as the drug never made it to testing and was shelved for years. For over half a century its euphoric effects remained completely unknown to humanity, until a chemist named Dr. Alexander Shulgin came along in the 1970s and noticed that MDMA's chemical structure resembled amphetamine and mescaline, two very powerful drugs. Being a skilled chemist, Shulgin synthesized his own MDMA and experimented on himself. As you might expect, he was pleased with the effects. So he shared it with his friend Dr. Leo Zeff, who distributed it to several hundred therapists in Northern California.

Through the early 1980s, these therapists experimented with giving their patients MDMA during sessions (of note, the drug was legal at this time). One such therapist, Dr. George R. Greer, published several case reports about his trials, noting that his patients could explore traumatic memories with less fear than usual. Now, decades later, modern research is suggesting that MDMA can be very effective for treating post-traumatic stress disorder (PTSD). Greer's work was an early hint at this.

These therapists also reported something else remarkable: Their patients felt extremely *empathic* and *connected with others* on MDMA. It made them more extroverted and social, while bringing on feelings of euphoria and happiness. Greer logged that MDMA gave his patients "effective skills for communicating feelings to family members." These effects were so unique that it necessitated a new term for drugs that enhance empathy: *empathogens*.

With these discoveries, a fascinating research question emerged: *How does MDMA enhance empathy in the brain?* If science could figure out how it was doing this, perhaps we could design new drugs that bring people together without all the intensity of MDMA—potentially for use in counseling and therapy. Unfortunately, science never got to solve this puzzle, as the U.S. Drug Enforcement Administration cracked down and made MDMA an illegal Schedule I drug in 1985. This promising area of research screeched to a halt and remained frozen for almost twenty years.

Then in 2004, MDMA reemerged when it was approved for the study of PTSD. Research on MDMA has been exploding ever since, with thousands of new papers published. But somehow the puzzle of the empathogen remained unsolved. While some studies showed that MDMA *does* increase empathy, there was little clarity about how it was doing this in the brain.

Surely, neuroscience could get to the bottom of this! While I never saw it coming, it turned out that I would be the person who led the study that answered this age-old question.

After finishing my PhD, I moved on to Stanford University for my postdoctoral fellowship, which is the last step in a scientist's training before they become a professor and run their own lab. It's basically the equivalent of a doctor doing a residency, but for scientists. I was fortunate to land a position working with a highly esteemed neuroscientist named Dr. Robert Malenka, who is responsible for uncovering a great deal of what we know about synaptic plasticity. As of this writing, his research papers have been cited a whopping 125,000 times. When I joined Dr. Malenka's lab, he had recently become interested in studying MDMA. Meanwhile, there happened to be another researcher in the lab studying empathy. This was none other than Dr. Monique Smith—the researcher from chapter 5 who discovered the social transfer of pain in mice!

Dr. Smith had seemingly discovered a form of empathy in mice, and given the lab's interest in studying MDMA, the question naturally arose: What would happen if she gave them MDMA? Would it *enhance* how much they took on one another's sensations?

If so, perhaps we could finally get to the bottom of MDMA's empathogenic effects by studying the brains of these mice. Dr. Smith ran the experiment, and it worked. Mice on MDMA were seemingly more sensitive to the pain of their friends, taking on their hypersensitivity for much longer than sober mice did! At long last, neuroscience was positioned to figure out what exactly empathogens do in the brain. However, at that point, an obstacle emerged. Dr. Smith was leaving Stanford to open her own lab in San Diego, leaving the project with an uncertain future.

I had been at Stanford for only a few months, and I was working on a totally different project, studying how social experiences in early life shape brain development. When Dr. Smith asked if I would like to take over the project, I was blown away. I thought back to when I had been sitting at my desk three thousand miles away in Buffalo, reading her paper in utter disbelief. Now, I had a chance to study this phenomenon directly. Even better, I would use it to study how MDMA enhanced empathy in the brain. How could I say no?

I got to work immediately. For my first experiment, I would place a mouse into a small arena with MDMA pulsing through its tiny vasculature, then leave it alone for fifteen minutes to allow the drug to set in. We used a fairly high dose of MDMA (15 milligrams per kilogram of body weight), and I could tell when it kicked in because the mouse would look sort of puffy. MDMA causes *piloerection* in mice—the equivalent of goose bumps—so their fur would stand up, making them look slightly larger and a bit spooked. After that fifteen-minute period, I would bring another mouse into the arena, one who had arthritic swelling in one of its paws. As is customary in mouse culture, the two mice would greet each other with some sniffing and chasing. Then I would leave the room and let them interact for ten minutes. After this brief hangout, I tested the sensitivity in their paws by prodding them with those thin bristles every few hours over the next couple of days. That way, I could compare the sensitivity of those MDMA-treated mice versus controls who were sober during their interaction with the arthritis mouse. So, were the MDMA-treated mice more empathic?

The effect was clear as day. MDMA-treated mice were *far more* hypersensitive than the controls; it wasn't even close. This was incredible to see with my own eyes, and there was a palpable sense that we were on the precipice of something exciting. After more than half a century of mystery, it was high time to finally figure out how MDMA was amplifying empathy in the brain.

There are 737 distinct regions in the mouse brain, and MDMA could have been acting in literally any of them. To narrow it down, we looked at which brain regions were *more* active in the MDMA mice while they interacted with the arthritis mouse, hoping this might offer a clue about where MDMA was acting. Interestingly, we found that MDMA increased the activity in the nucleus accumbens, that M&M-like brain area involved in social reward. This made perfect sense, considering the region's role in other social functions like social reward. Not to mention, Dr. Smith had previously found that the nucleus accumbens is activated when mice take on one another's pain, even without MDMA on board. If MDMA was increasing the activity there, perhaps the drug was simply turning up the dial on the brain's existing empathy systems?

Our next experiment seemed to confirm this. When we injected MDMA *only* into the nucleus accumbens, but nowhere else in the brain or body, the mice showed a huge boost in empathy. It seemed we had found the key brain area where MDMA was acting, but *what was it doing*?

MDMA hits three neurotransmitters in the brain: serotonin, dopamine, and norepinephrine. More specifically, it hijacks the machinery that transports these signals between neurons, causing all three to be released at way higher levels than usual. This may explain how MDMA *enhances* so much of human experience. As I mentioned at the beginning of the chapter, everything we experience is simply chemicals in the brain. When those chemical signals are spurted out in abnormally high volumes, we may experience inordinate levels of things like connectedness and empathy. Perhaps one of these chemicals was the key to empathy? Like detectives trying to nail a suspect, we set out to determine which was responsible. Luckily, we had a very strong clue.

A few years earlier, another Stanford researcher named Dr. Boris Heifets had found that MDMA makes mice much more interested in hanging out. If given the choice between spending time with another mouse or with a toy, they will much prefer the social option if they're high on ecstasy. Impressively, Dr. Heifets tracked this down to a specific signal in the brain: *serotonin release in the nucleus accumbens*. If we think back to what we learned in chapter 1, this explains a lot. Serotonin in the nucleus accumbens is a signal for *social reward in the brain*; so if MDMA causes a mother lode of serotonin to be dumped there, it might create an exaggerated sense of social pleasure. With this in mind, we suspected that serotonin might *also* be responsible for MDMA's effects on empathy—and we suspected right.

I set up a final experiment, but this time there would be no MDMA involved at all. Instead, I used optogenetics to stimulate serotonin release in the nucleus accumbens. In other words, I was cranking open the valve that controls serotonin release there, causing it to dump like mad. If serotonin was really the key to MDMA's effects, then serotonin alone should produce the same results . . . and it sure did. With their serotonin stimulated, the mice became much more empathic. What's more, we also found that this could *restore* empathy in mice carrying a gene mutation associated with autism spectrum disorder. These mice hardly took on one another's pain at all when sober, but when I stimulated serotonin release in the nucleus accumbens (or gave them MDMA), they showed a heightened sensitivity to one another's states.

At long last, the puzzle seemed to be solved. MDMA seems to enable a totally unprecedented level of empathy by driving serotonin levels to newfound heights—at least in mice.

While this remains to be confirmed in humans, there's no doubt that MDMA enhances our empathy. In 2023, a news story broke about an American man named Brendan who was a white supremacist and the leader of a white nationalist group. Brendan happened to take part in a study involving MDMA, and afterward, his views on race changed dramatically. He left a note for the researchers that read: "This experience has helped me sort out a debilitating personal issue." When asked for more information, he clarified that he no longer identified with his racist beliefs. "Love is the most important thing," he told them. Amazingly, his single experience with MDMA had made him second-guess his views on race.

Recall that in chapter 4, I proposed that we all exist somewhere on a spectrum of empathy. I'd like to think that Brendan was simply low on the slider of empathy *for different races*. Perhaps for the first time in his life, MDMA pushed his slider farther than it had ever gone and opened a window of perspective. Maybe the serotonin gushing into his nucleus accumbens blew out the boundaries of his typical empathy and let him feel the emotions of people he had rashly abandoned. With such an abundance of empathy, the brain almost has no choice but to dismiss the barriers that divide us. We can appreciate all perspectives, regardless of our self-other overlap.

I can't help but feel amazed and perplexed that a drug can jostle the brain's social systems in a way that alters our social perceptions. As complex as our brains and relationships are, a single compound can flip all that upside down and make us care for one another. We have the chemical capacity within us, and certain drugs can push those systems into overdrive. Because just like anything else, empathy is a chemical signal in the brain.

## HOW PSILOCYBIN WASHES AWAY THE "SELF"

The discovery that MDMA acts through serotonin is a significant one. It doesn't just explain how MDMA works; it also suggests that serotonin is central to empathy. In our quest to understand the brain systems that bring us together, this points our flashlight in the right direction.

It also opens another question: If serotonin is really that important, then other drugs that target serotonin should influence our interactions, right? To answer this, we might look at something like *psilocybin*.

Psilocybin is the compound that makes psychedelic mushrooms psychedelic. After entering the body, it gets metabolized into a compound called *psilocin*, which can bind directly to the brain's serotonin receptors and activate them, with a special affinity for the serotonin 2A receptor. When psilocybin is in your system, your brain thinks that a mother lode of serotonin has been washed over its 2A receptors, causing them to be activated at levels far beyond what would happen normally. This is thought to be responsible for the psychedelic trip. But if psilocybin activates serotonin receptors, does it enhance social connection in some way?

You bet it does.

Psilocybin is known for a few trademark effects like distorting the appearance of colors and textures, generating spiritual experiences, and *creating a sense of oneness with everything*. I'm particularly interested in this last one. Friends have told me about their experiences on psilocybin, describing this feeling of profound unity and connectedness. As they leaned against the trunk of a tree, they realized there was no difference between themselves and the tree. A toad skipped by, and they felt a sense of harmony with it. It became clear that all living creatures are somehow one and the same—intrinsically unified on the team of "planet Earth." It sounds profound.

Science has a name for this experience: It's called *ego dissolution* or ego death. The term comes from the idea that one's sense of self (or ego) dissolves, giving way to the sensation of being one with the universe. It usually happens only at high doses of psilocybin, and it probably means that the brain systems that keep track of the self get mixed up when the serotonin 2A receptor is unnaturally roused. But personally, I can't help but focus on the social nature of ego dissolution. To become indistinguishably blended into a swirl of others is intensely *communal*. It's unlike the brain's typical habits of finding ways to divide the self from others. For most people, even the most beautiful tree or the most humanlike toad would never be considered a member of their in-group. Disregarding the boundary between self and other may be the most powerful social experience you can have, as it forces the brain to pocket any perceived differences. You can't judge another person for their differences when there is no such thing as another person. That person is simply *you*.

With this sense of oneness taking hold, the brain seems to process social interactions differently. When people high on psilocybin were left out of the passing rotation in Cyberball, they reported feeling *less excluded* than sober people. Apparently the sense of unity created by psilocybin is so profound that it stands tall even in the face of plain neglect. What's more, the people on psilocybin showed less activity in the dorsal anterior cingulate cortex, a region responsible for processing social pain.

Psilocybin has also been shown to enhance emotional empathy, which should come as no surprise considering this tendency to wash away the distinction between self and other. Recall that the brain is prone to experience more empathy for people we see greater self-other overlap with. When the boundary between self and other is blurred (or nonexistent), it's only natural that empathy would spike.

Perhaps for the same reason, psilocybin has been linked to more altruistic behaviors. In a study from King's College London, subjects were offered twenty euros, but with a catch: They had to split them with another participant. Sometimes the splits were favorable (like an 80/20 split in the subject's favor), while others were downright unfair (like 10/90 or 20/80). Interestingly, the participants on psilocybin were much more likely to accept offers that disadvantaged them. This is a distinctly altruistic decision—one that helps the other person at their own expense. The researchers proposed that psilocybin drove the subjects to value their *relationship* with the other player more than the *financial reward*, and I can't say I disagree. When we feel more connected with others, we may be more inclined to act selflessly to benefit them. This shift in behavior might come down to a swath of microscopic serotonin receptors being tickled just the right way in a specific nucleus of the brain.

Once again, more evidence points to serotonin as a key player in the brain's social systems. Drugs that stimulate serotonin signaling not only enhance social interest and empathy (MDMA), but also enhance unity and altruism, while impairing social exclusion processing (psilocybin). However, not so many people take these drugs, and there is another, much more commonly used drug that acts directly and specifically on serotonin signaling.

## HOW ANTIDEPRESSANTS INFLUENCE THE SOCIAL BRAIN

Selective serotonin reuptake inhibitors (SSRIs) are some of the most highly prescribed drugs in the world. They're most often used to treat depression but are also prescribed for other conditions like anxiety and OCD. As you can tell from the name, SSRIs act on serotonin in the brain. But what the hell does *selective serotonin reuptake inhibitor* mean?

When neurons send neurotransmitters to one another, it's a bit like sending mail. One neuron spurts out some neurotransmitter (which we might think of like a letter in an envelope), which floats between the two cells and lands in a receptor (like a mailbox). What's more, neurons have a built-in return-to-sender feature. After a neurotransmitter like serotonin gets released, it often gets slurped back up by the neuron that released it. This process is called *reuptake*. SSRIs *prevent* the reuptake of serotonin (hence their name), which causes that extra serotonin to stay floating around in the space between the neurons. Consequently, the receiving cell ends up receiving more serotonin signals over time. Therefore, SSRIs *enhance* serotonin signaling.

Considering that SSRIs act on serotonin, we might suspect that they would promote social connection like psilocybin or MDMA. Indeed, some evidence suggests this to be true.

In one clinical trial, patients with depression who took SSRIs became more extroverted, while those who took a placebo showed no such changes. Now you might be thinking, *If someone is relieved of their depression, couldn't that make them more outgoing?* The researchers thought of this, too, and ran an analysis to check for it. It turned out the patients on SSRIs became more extroverted regardless of how well the drug treated their depression. Perhaps this was because of its effects on serotonin?

In another study, SSRIs were shown to reduce hostility and boost affiliative behavior. After taking either an SSRI or a placebo for four weeks, subjects were paired up to solve a puzzle together, but with an annoying twist: Only one person could touch the puzzle pieces at a time. As irritating as this may have been, the subjects on SSRIs were more likely to make helpful suggestions and issue fewer commands than those in the placebo group. Even more convincing, people who had higher concentrations of SSRIs in their bloodstream were more likely to engage in these friendly behaviors. This means that people on SSRIs either behave more considerately, or they can better tolerate the hellish experience of watching a stranger solve a puzzle. Hard to tell.

Evidence from another study would suggest the former: that SSRIs specifically bring out our positive social nature. The subjects were presented with various challenging moral dilemmas. For example, imagine you're standing at a train station, and the train is barreling toward a group of five people on the tracks. You can save them by pulling a lever that diverts the train to a different track, but there is one person on that track who will be killed by your action. Will you pull the lever?

Tough decision, I know. Now let's add a layer. Same situation: A train is speeding toward a group of five, but this time you can save them through a different method. You can *push* a large, heavy person in front of the train to stop it. You are sacrificing the life of that person to save five, but you must physically push them yourself.

This second dilemma is much more challenging because you are directly involved in the act. It's far more personal to *shove* someone in front of a train than to pull a lever and watch the tragedy from afar. Interestingly, people taking SSRIs rated these more personal acts as *less acceptable* than those given a placebo. Somehow the drug made the idea of personally harming someone more repulsive and objectionable.

As time goes on, neuroscience will surely discover more ways that drugs influence our social lives—especially as new drugs continue to be introduced through medicine. In the meantime, it's important to recognize how the substances we ingest may be altering how we process social information. The decisions we make in social settings may not always be our own. Rather, they may be predetermined—or at least influenced—by the chemical forces meddling in our minds. It's a good reason to pause in social settings and consider the question *What influences am I under?*

## Key Takeaways

1. Since the brain translates the outer world into electrochemical signals, drugs that influence these chemical systems can alter our experiences and perceptions.

2. Alcohol suppresses brain activity by acting on GABA receptors. It has been shown to blunt social information processing.

3. Painkillers like acetaminophen can reduce social pain and blunt empathy for both positive and negative experiences.

4. Anxiety medications may limit acts of compassion by reducing how much distress we feel from seeing others struggle.

5. MDMA enhances empathy through serotonin signaling, specifically in the nucleus accumbens.

6. Psilocybin, which also acts on the serotonin system, can increase feelings of unity and make people feel less excluded, while enhancing empathy.

7. SSRIs, which also act on serotonin, can produce prosocial effects. This collectively suggests that modulating serotonin is key to influencing social behavior.

To view the references cited in this chapter, please visit benrein .com/book.

# 10

————— ◦ —————

# HUMANITY'S
# BEST FRIEND

Why Loving Dogs Is Good for Your Brain

In a book about the value of social connection, it would be a shame to leave out the relationships we share with dogs. They have been man's best friend for a very long time, and they deserve some respect. Not just because they're cute and wonderful and fill us with joy, but because they, too, offer tremendous value for our health and well-being. We can extract *so much* from these connections, and it's no wonder why. As you'll soon learn, our brains treat them much like human bonds. It's something I felt really deserved a spotlight in this book.

Growing up, I had the fortunate experience of being surrounded by animals. I wasn't raised on a farm—just one of those households with way too many pets. It all started with a cat named Missy. My parents adopted her when they first got married but weren't quite ready for children. By the time my sister and I were reaching middle school, Missy was getting old. It was time for another pet—but my parents overcorrected.

For mysterious reasons, they adopted three dogs and two cats all within the next year. Suddenly we lived in a home where you needed to be very careful where you stepped. There was also an occasional guest appearance by a goldfish, hamster, hermit crab, betta fish, frog, or some other creature. We enjoyed a wild and wonderful childhood full of many little visitors.

I don't know what my parents were thinking, but I'm grateful for their impulsiveness. Their arguably poor decision-making gifted me the splendid experience of growing up alongside animals. It allowed me to appreciate from a young age that dogs and cats are sophisticated beings with impressive social capabilities. The friendships I shared with them didn't feel terribly different from my friendships at school. Each pet had their own unique quirks and little flairs of personality that contributed meaningfully to our relationship. I truly believed—and still do—that I understood them and the messages they shared with me. I'm not talking about things like "I am going to meow in your face as loud as I can until you give me a can of tuna." It was deeper than that. They shared complex messages with me, like "I'm not feeling well today. Could you do your homework on the couch so I can be near you?" I could sense when they had something to share, and I would parse out the meaning by carefully watching their behavior and expressions.

It was heartbreaking and tragic when we eventually lost these beloved pets. Since my parents got those three dogs and two cats all within a short window, they all reached old age together and passed within a short period of a year or two. We mourned them as we would grieve the loss of any family member. Our home had shrunk from a stout nine members to just four—all of which were human. The house felt cold and empty. Some of its loudest personalities and central figures had moved on.

Despite the pain and anguish we experienced, there's no doubt it was all worth it. Those animals left a deep and permanent impression upon me, as they literally participated in the shaping of my brain. They were the ones that greeted me with wagging tails when I returned from my first day of school or licked up my tears when I learned some of my first life lessons. While I grew up and discovered that life is full of challenges, their love was one of the few things that always held constant. My memory of them still burns bright today, a flame that will probably flicker on until the electrical activity one day leaves my brain.

If you've ever had a pet, I'm sure you can relate. Surely hundreds of millions of people—or perhaps billions—have shared a similarly deep bond with a dog. But if you step back—like, *way* back—this is quite astonishing and unexpected. I mean, what do humans and dogs really have in common? We don't even have the same number of legs. But somehow we find ourselves sharing beds and kisses. How did this unlikely partnership come to be?

The bond between humans and dogs goes a long way back. In northern Israel, the fossilized remains of a human skeleton and a puppy were found buried together beneath a *12,000-year-old home*, the small animal cradled in the human's arms. This means that in roughly 10,000 BC, dogs were already an integral part of human life. Astonishingly, genetic evidence suggests an even longer history. When scientists compare the dog genome with the wolf genome to figure out when dogs split off and were domesticated by humans, they find hints that it happened between 27,000 and 40,000 years ago.

Yes, this is true. Not only have we been BFFs with dogs for that long, but the sweet little pups who lick our hands and bare their bellies for scratches did in fact evolve from wolves. There are a few theories about why this happened. One proposes that some wolves got nutrients from eating human poop and garbage, so they slowly grew accustomed to living near humans. This gave an evolutionary advantage to the wolves who were naturally less aggressive toward humans, because if they were nice to us, they got an unlimited supply of human excrement. Perhaps this is why some dogs eat their own poop. It might be an artifact of their ancient history, a behavior that is no longer useful in the modern world but is still so deeply ingrained in their origins that it remains unerasable from their biology. Or maybe they're just nasty.

A second theory is that during severe cold, our hunter-gatherer ancestors may have had extra meat they couldn't eat, because humans can eat only so much meat before we have issues like diarrhea. They may have shared this excess with wolves in exchange for protection and some assistance in hunting. Over time, this led to the gradual domestication of wolves. This seems somewhat feasible to me, but I do wonder why the wolves wouldn't just kill our ancestors and take all their meat.

There's also a third possibility: that humans simply kidnapped wolf puppies and raised them into companions or hunting partners. I view this as the least likely theory. Not only would it be extremely risky and dangerous to capture a wolf puppy, but it would also be really time consuming and challenging to raise one, especially while constantly being hunted by its parents. It makes little sense to devote resources to this in an era where the top priorities were to find food and raise our own young.

Regardless of the exact cause, humans and wolves were shoved together, and it must have worked well or it wouldn't have lasted all this time. During this period, both humans and dogs faced extreme threats to their survival, like inclement weather, food scarcity, and predation. In a world of danger, humans and dogs must have been better at navigating these challenges when they worked together. Thus humans gained a best friend.

Then, something amazing happened. As we teamed up and took on the world together, our genes began to *converge*. When scientists compare the dog genome with the human genome, they find that our genes have changed in many of the same ways throughout history. At first, that might sound like we were somehow becoming more doglike over time, but it actually reflects something very different: that we were *facing the same challenges*. As we were exposed to the same environmental pressures over thousands of years, our genes had to take on the same modifications to allow us to adapt and survive. Just think about that: Dogs have been man's best friend for such a long time that our DNA has literally been shaped in the same ways. It's like our bond is inscribed in our genetic code. Those common changes in our genes are like shared scars, emblems revealing that we braved the great storms of ancient Earth together and made it out as one.

## ARE DOGS JUST HAIRY BABIES?

Back in the 1960s, psychologist Dr. Mary Ainsworth designed a landmark procedure called the *strange situation test*. It was meant to identify attachment styles in infants by putting them into . . . well, a strange situation.

The child would enter a room with their mother and be let loose to explore and play with a scattering of toys. After a few minutes, a stranger would join them and try to play with the child. Soon after, the mother would exit the room, leaving the child and stranger alone. Is it getting awkward in here? Luckily, the mother would return after a few minutes.

Ainsworth believed that the way a child behaves in this situation can reveal a lot about their attachment style. Most babies (about 70 percent) display a *secure* attachment style: They become distressed when their mother leaves and are skeptical about playing with the stranger in her absence. When she returns, the child seems happy and becomes more willing to play with the stranger. Children with a secure attachment style thus see their mother as a secure base from which they can safely explore. Mom is a beacon of safety.

Another 15 percent of infants demonstrate an *anxious-avoidant* attachment style. They show a sort of indifference to their mother, reacting very little when she leaves or returns, and play freely with the stranger while she's gone.

The remaining 15 percent show an *anxious-resistant* attachment style. They become extremely distressed when Mom leaves and generally avoid the stranger in her absence. But then when the mother returns, the child becomes resentful. They might resist her contact and seem to hold a grudge against her for leaving them with this weird, freaky stranger. How could you do that to me, Mom?!

You may be wondering why I'm sharing all this in a chapter about dogs. Well, a few decades after Ainsworth's studies, researchers wondered if *dogs* show these attachment styles for their caregivers, too. Considering that human/dog relationships tend to resemble parent/child relationships, I could see this idea having legs. But the question is, what did the dogs think?

The results are sure to warm the heart of any dog lover. When dogs were put in the strange situation test, they showed a secure attachment style to their owner just like most infants do. They were much more likely to explore an area when they were accompanied by their owner, and less likely to play with a stranger when the owner left. It seems that dogs view their owner as most infants view their mother: a secure base offering safety and comfort.

This raises an incredibly important scientific question: **Are dogs just hairy babies?**

A surprising amount of neuroscience research says yes. Not that dogs are literally small children in disguise, but that the relationship between a dog and its caregiver is—on a neurological level—akin to a parent/child relationship.

Remember in chapter 7 we learned that oxytocin goes up when parents simply look at their child? Well, the same happens with our pups. That's right: When dogs and their caregivers look at each other, they both show a significant rise in oxytocin levels. What's more, when dogs are given *more* oxytocin (for instance, by researchers in a lab), they spend even more time looking at their owners. Amazingly, this increased eye contact seems to set off a positive feedback loop, causing the *owner's* oxytocin levels to rise. It's like staring at your dog creates a sort of oxytocin hyperloop! But why?

Recall the role that oxytocin plays in the parent/child relationship: Both parent and child produce heaps of oxytocin, which acts like a sort of glue forming a rock-solid bond between them. This is the work of evolution: By drawing parents tight to their child, evolution ensures that the baby will be protected and safe. Now, let's consider dogs. Just like babies, our dogs benefit *a lot* from cozying up to us. Some thirty or forty thousand years ago, dogs discovered that teaming up with humans offers some survival advantage—probably involving free food (not much has changed, huh?). Because of this, dogs may or may not have evolved to hijack our brains' oxytocin system to gain our protection and care. What evidence is there for this? Well, when wolves stare at humans, they show no such rise in oxytocin (and neither do the humans), suggesting that this oxytocin system evolved during domestication. This oxytocin may also be why we feel rewarded by the presence of dogs and get attached to them.

This isn't the only trick dogs have up their sleeves; they are packed with tons of features that hijack our vulnerable brains and make us love them, like *puppy eyes*. Research shows that dogs in kennels who raise their inner eyebrows more (creating that irresistible puppy dog look) are adopted much faster. This is literally a real-world example of how being cute is evolutionarily advantageous for dogs. The pups who were better at pulling on human heartstrings were legitimately more likely to be brought into homes and taken care of. This, too, is a unique feature that dogs evolved to connect with us. Wolves physically can't do puppy eyes because about thirty-three thousand years ago, modern dogs evolved a unique muscle called the *levator anguli oculi medialis*, which lets them raise their inner eyebrows and create that irresistible expression. Wolves simply don't have this muscle and therefore can't emote as intensely as dogs to win us over.

Another feature that helps dogs connect with us is *eye watering*. If you have a dog, look at their eyes the next time you come home from work or otherwise reunite after at least a few hours apart. If they have a glistening, wet appearance, it may be a sign of their love for you. When dogs are reunited with their owner after several hours, they produce more tears in their eyes than when reunited with a stranger. But why? It again seems to be for the same reason: because it makes them cuter. Indeed, people rate teary-eyed dogs more positively. This, too, seems to be driven by oxytocin: When a solution containing oxytocin is applied to dogs' eyes, they produce more tears than when they received the same solution without oxytocin.

It may seem like this is all just a big scheme by dogs to make us love them. When your dog looks at you, *you* produce oxytocin, and *they* produce oxytocin, which makes their eyes get all wet and beady and cute, which makes you love them *even more . . .* and suddenly you're on a slippery slope to buying them expensive treats and walking them multiple times a day.

The fact that dogs evolved these traits shows how valuable we are to them. They wouldn't have adapted these new features just to bond with us unless we greatly improved their survival chances. On the other side of the equation, we benefit, too. As we learned in chapter 1, our brains are packed with social reward systems because we survive best in groups. Dogs presumably hit on these systems because being around them increases our survival chances. Also, this isn't just some evil plot by dogs—they feel the love, too. The fact that they release oxytocin when they look at us—which is deeply linked to love in the brain—means they probably associate us with social reward. I would even venture that dogs almost certainly love us. Perhaps they view us as their parents, just as we see them as our "fur babies."

Considering that dogs have piggybacked on the brain's systems we use for bonding with our children, perhaps we can take advantage of that for our health. In those moments when you're looking in your dog's eyes or holding them close, you may be producing that oh-so-salubrious oxytocin, which may carry health advantages. How lucky are we to have the privilege of these beasts? We have the chance to surround ourselves with animals that boost our health the same way that time with our children does, and I believe that makes dogs a valuable element in one's social diet. Of course, we shouldn't replace all human contact with time spent with dogs, but our beloved pups may help buffer against the effects of loneliness in the absence of interaction. Especially for isolated older folks who are at higher risk of health consequences, a dog may be tremendously valuable. If Grandma or Grandpa can take care of a dog, it might be a great way to restore some of that diminishing social contact they experience later in life.

I do want to acknowledge that other types of pets can also be great for you. Cats can help buffer against stress, and owning a cat is associated with reduced likelihood of cardiovascular death. However, dogs may carry slightly stronger effects, which would make sense considering our unique, long-lasting partnership with them. For example, one study of more than three thousand military veterans found that dog owners showed lower blood pressure and cholesterol, along with lower rates of heart disease and diabetes, while cat owners did not. Dogs happen to be the most well-studied human companion, but I think having any animal partner can be healthy. As a real-life validation of this idea, children with pets (including dogs, cats, fish, birds, and others) had much lower rates of disordered sleep and hyperactivity during the stressful COVID-19 pandemic. Any animal may be good company, but given the evolutionary pressures that drove humans and dogs together many millennia ago, our brains may be wired so that dogs produce the largest benefits of all.

With that said, it's time we dive a bit deeper into how dogs help regulate human health. After all, I wouldn't have devoted an entire chapter to them if the effects weren't for real. It's time we explore exactly what we stand to gain from hanging out with our hairy little babies.

## BARK THERAPY

Therapy dogs have had more impact on humanity than you can shake a stick at (pun intended). Every day, canines are brought into hospitals and nursing homes and onto college campuses to help people feel better. These efforts aren't just cute and fun; they really work. For instance, on-campus interactions with animals have been shown to improve students' mental health.

Service dogs can also provide medical assistance for a plethora of health conditions. They offer sensory support to those who have difficulty seeing or hearing. They retrieve objects and open doors for those with motor conditions like cerebral palsy and multiple sclerosis. They can find lost items or serve as a guide for dementia patients. For people with narcolepsy, a sleep disorder characterized by sudden sleep onset, service dogs can detect oncoming sleep events and alert their handler in advance. In similar fashion, dogs can warn epilepsy patients about impending seizures.

How is this even possible? Can dogs see the future? It turns out, dogs have a different superpower: They can detect signals in human *sweat* that tell them a seizure is coming. In a 2021 study, researchers harvested sweat from epilepsy patients before, during, and after seizures. Then they applied this sweat to non-epileptics in the presence of their dogs. When the dogs detected the seizure-associated sweat on their caregiver, they inexplicably became more attentive to them, spending more time near the caregiver and engaging in more eye contact. Importantly, these were *untrained* dogs interacting with their non-epileptic owners. This suggests that dogs may naturally perceive a physiological distress signal in human sweat.

Of course, we also can't forget about the countless emotional support animals that provide comfort and companionship to humans. Dogs have *proven* calming effects on the human nervous system; interacting with them has been scientifically shown to reduce anxiety and boost mood. But what's really happening in the body?

To answer this, one study measured physiological signals in subjects before and after interacting with a random dog—talking softly to it, petting it, playing with it, and scratching its ears. Not a bad experiment to be selected for! Amazingly, interacting with the

dog reduced their blood pressure, increased their oxytocin and dopamine levels, and more than doubled their levels of endorphins.
Another study (which I absolutely must mention was led by a researcher named Dr. Sandra *Barker*) found that interacting with a
therapy dog could reduce cortisol, the stress hormone. Dogs seem
to drive several health-giving effects in the human body, which
may explain why they make such valuable emotional support animals.

Now, I've been holding out on my absolute favorite part of
these studies. The researchers also measured some metrics from
the *dogs,* finding that playing with a human caused their blood
pressure to drop slightly and drove up their levels of endorphins,
oxytocin, and dopamine. Maybe that's why they're always so eager
to play with us!

Being around a dog isn't just good in the short term; it also has
long-term benefits. Owning a pet (either a dog or cat) has been associated with a 19 percent lower risk of death from cardiovascular
disease, and pet owners also show lower heart rates and lower
blood pressure. How can this be?

One explanation is that people with pets (especially dogs) are
simply more likely to walk outside. It's true, people with dogs have
higher daily step counts, which surely benefits one's heart health.
However, the dog itself seems to be important, too. One study
found that when healthy seniors walked *with* a dog, it produced
more favorable indicators of cardiac activity than when they walked
*without* a dog. That could be because dogs push us (or more appro-
priately, *pull* us) to new athletic heights, but it also might be that
dogs simply calm the human body. In that same study, the seniors
showed greater signs of parasympathetic nervous system activity
(think: rest and digest) while at home with the dog versus without
the dog. This is consistent with those other studies showing that
dogs improve heart rate and blood pressure. It seems that being
around a dog is like putting an ice pack on the nervous system,
which could be why living with one produces long-term benefits
on health.

These soothing effects could also mean that being around dogs might help us manage stressful situations. Indeed, studies show that having a pet nearby when stress strikes can be protective. At the University at Buffalo, researchers had subjects engage in two stressful activities: a mental math task, where they had to quickly perform subtractions from a large number; and a cold challenge, where they placed their hand into an ice bath for two minutes. The subjects faced these challenges alone and with their spouse, a friend, or their pet (both cats and dogs were included).

Amazingly, when the subject's pet was present, their heart rate and blood pressure were lower during both challenges. What can we learn from this? Well, if you happen to receive a stressful phone call or bad news in the presence of your pet, perhaps you'll be less likely to show a pronounced cardiovascular spike—and that could protect you against things like heart attacks.

Clearly, the benefits of dogs (and cats) go well beyond the simple company they provide. These beings offer much of the same value as human contact—a fact we should acknowledge and appreciate. Maybe instead of turning to a stiff drink to ease your nerves after a long day, you can give your dog a scratch and look in their eyes. Your brain will thank you—and theirs will, too.

## RAISED BY WOLVES

At the beginning of this chapter, I described my experience growing up alongside dogs and cats. When I said that my pets taught me a lot, I really meant it. They helped me learn and understand social dynamics. Of course, this was only my anecdote, but there is now scientific evidence showing that dogs can enhance social skills in children and offer developmental advantages.

For example, preschoolers who come from families with a dog show lower rates of conduct problems such as exhibiting temper tantrums, fighting with other children, and arguing with adults. Having a dog is also associated with prosocial behaviors (like helping others who are hurt, being considerate, and sharing) and having fewer peer problems (like not getting along with other kids). Caregivers also tend to rate children higher in emotional expression if they had pets as toddlers.

Why is this? Personally, I believe that having pets in early life helps children learn empathy. As we discussed earlier, humans aren't born with empathy—we learn it through experience. When a child is given a pet, they must learn to understand the animal by interpreting its social cues. Think about it: Is there really *any* better way to learn empathy than searching within the mind of a nonverbal being and sensing their internal state? Indeed, kids with a stronger attachment to a pet show higher levels of empathy, and adults who grew up in cultures with extensive exposure to dogs are better at recognizing emotions in dog faces. Kids with pets are also more likely to feel sad when seeing others in distress and to show greater compassion in responding to others. *Paging all parents*: If your child has been begging for a puppy, don't shoot down the idea so quickly. Having a pet facilitates the development of perspective-taking and compassion; furthermore, it provides valuable *parental* experience, as children may take on the responsibility of caring for a being that is more vulnerable than they themselves. For a budding human, learning these skills early in childhood is invaluable.

⊚

The fact that kids can learn empathy from being around dogs implies something meaningful: Humans inherently care about dog emotions. You wouldn't expect a kid to learn empathy from sitting in a room full of fruit flies, because we generally don't empathize with those critters. In contrast, we care deeply about how dogs feel. This makes sense considering that dogs have hijacked the brain's bonding systems. This deep, oxytocin-driven attachment we have to dogs is a breeding ground for empathy. In one study, college students read news articles (which luckily were fake) describing a victim being assaulted with a baseball bat. The victim was either an adult human, a baby, an adult dog, or a puppy. Remarkably, people reported more empathy for the puppy or the adult dog than they did for the adult human. Perhaps this has something to do with the way people view dogs: like children, or "fur babies." Dogs are often thought of as innocent and pure, deserving of protection and care. Maybe that's because they've commandeered our brain's systems for parent/child bonding.

In any case, there is another parallel here. Just like those dolphins who were more generous in front of their children, humans tend to be on their best behavior around dogs. Really, dogs bring out the best in us. Have you ever felt more inclined to give someone money on the street because they had a dog with them? You're not alone. In one study, an actor walked the streets of France with or without a dog, asking strangers for money. When the actor was accompanied by a dog, more than twice as many people coughed up some cash, and they gave 39 percent more on average. When he spilled some change on the ground, people helped him pick it up 87.5 percent of the time when he was with a dog, versus only 57.5 percent when he was alone.

Dogs seem to tickle something within us, stirring up positive feelings that can rub off on the folks around them. People accompanied by a dog are perceived as more likable, and research even shows that men may be judged more favorably when they have a dog. When a handsome twenty-year-old actor asked women on the street for their phone number, only 9.2 percent of women obliged. When he had a dog with him, his success rate jumped to 28.3 percent.

I think it's a blessing that dogs bring out the best in us. How lucky is it that our children can learn empathy at home, and we can treat others with more respect *just because* our ancient ancestors some forty thousand years ago were more likely to survive with a canine partner? It's something that deserves to be celebrated. Perhaps it's also an opportunity to ask ourselves, why do we sometimes fail to treat one another with this same respect?

Just like dogs helped us survive through ancient challenges, we've also been there for one another in just the same way. Is it possible for us to return to our primitive roots, relying on one another for comfort and safety? In a world where our differences tend to jump out and become more salient than our commonalities, maybe we can try to see the purity of a dog in one another. Because let's face it, we're all just animals that want some love and attention.

◯◯

I hope this chapter might motivate you to spend time with your pets, perhaps to reap the associated health benefits or to bestow *them* with these same benefits. However, there's another reason that's probably even better: because you are their everything.

The great sadness of domestic pets (cats and dogs and fish and birds and all the rest) is that they spend almost their entire existence *bored*. It pains me to see my dog, Zoey, slog through her days while I work. She cycles from room to room, napping on chairs and couches and floors and rugs. The most exciting thing she might encounter is a stimulating sight out the window. Maybe she enjoys it, but to me, it looks like a drag.

The one exception to this lifestyle is the time we choose to devote to them. I can only imagine that it means everything to them when we jump on the floor spontaneously, grabbing their favorite toy and tugging or bouncing or chasing. Those moments *are* everything to them. It's what they live for. Without the walks, meals, and playtime, there is sadly little other substance to their lives.

Meanwhile, we humans are inundated with stimulation and are always looking for more. A single task is often insufficiently engaging for us. While we cook dinner, we open YouTube videos or organize the pantry. As we wait for the water to boil, we scurry into the other room to turn on the TV. We keep two or three things running at once, somehow *always* feeling inadequately entertained.

But for a moment, imagine the delight your pets would find in preparing themselves a meal. Cooking pasta or even making a sandwich might be the highlight of their entire life. To us, it's merely an inconvenience—something pulling us away from the many attractive options we have for spending our time. Despite all we have in common with dogs, we lead very different lives and use our attention in very different ways. Perhaps we should be a little bit more thoughtful about how we use ours, and grateful for the many different choices we have for distributing it. And when you think of this, consider distributing a little bit more to your pet. Because you might be the only thing that they have in life to pay attention to.

And you're certainly their favorite.

## Key Takeaways

1. Dogs and humans have a long history of companionship, as partnering together supported the survival of both species.

2. The dog/caregiver relationship strongly resembles the parent/child relationship. Dogs show similar attachment styles to their caregiver as infants do to their mother, and oxytocin plays a major role in bonding humans and dogs.

3. Having pets can be tremendously healthy. Interacting with a dog lowers blood pressure, drives dopamine release, and lowers cortisol. In the long term, having a dog can buffer against stress and reduce risk of cardiovascular disease.

4. Exposure to pets in early life can support the development of social skills in children.

To view the references cited in this chapter, please visit benrein .com/book.

# MOVING FORWARD, TOGETHER

———— ○ ————

A few Septembers ago, I visited the beautiful city of Chicago. I was there to film an interview for a documentary about the neuroscience of sleep, and it was a very stormy morning when I left my hotel. Rain was already pooling in the streets when my Uber pulled up. As I leaned into the car, a spray of droplets showered down on me, wetting my suit. The car sped ahead, and I sat worrying in the back seat about how the interview would go and whether I'd be able to get dry.

My driver, Fernando, was a young guy—probably in his early thirties. He asked where I was headed, but otherwise it was a quiet ride. I was staring out the window preparing for my interview when the car unexpectedly slowed to a stop. Something seemed wrong.

"Is this . . . it, sir?" Fernando asked.

I looked out the window. Before me were a few dilapidated brick buildings, with train tracks looming overhead. This didn't seem to be a film studio. It was an old train station, at best. My Uber app showed that we had arrived at my destination.

My wet suit was quickly becoming the least of my problems. I started to search for the email from the producer, but Fernando already had his phone out. "Oh, it's just up a bit more!" he announced, showing me a waypoint on his map 1.2 miles ahead. The Uber app had sent him to the wrong address, but his phone's map knew where to go. "It happens sometimes," he said. "No worries."

I felt guilty that Fernando had to drive this extra distance that I hadn't paid for and almost offered to walk it. Then I remembered that it was pouring, and I imagined myself showing up to the interview drenched. I felt grateful for his generosity. As we drove on, Fernando and I began tuning into each other. Suddenly, this was more than a typical Uber ride; it was a voyage that we had been randomly paired together for. We both wondered how the mystery would end. Would we find the studio? I explained that I was headed to an interview, and he asked about my research. The conversation was lively. In just a few minutes, we learned quite a bit about each other.

We were deep in conversation when we pulled up to the real address, a large building with multiple entrances. I was now running a few minutes late and was back to worrying about the interview. I thanked Fernando profusely and started to get out, but he stopped me.

"I don't think this is the right entrance," he said. I squinted to make out a small sign printed on the door. He cranked the wheel and pulled the car around to the other side of the building, carefully reading the sign on that door. I glanced nervously at my watch.

"I want to make sure it's the right door," he explained. Then I realized that he genuinely didn't want me to get wet before the interview. He drove me around the building twice, ultimately pulling the car as close as he could get it to what seemed to be the right entrance. I thanked him again—genuinely, from the heart—and beelined through the rain into the building. I had made it safely to the right place, and it was indeed the correct door.

I never saw Fernando again, but I am still touched by his kindness today. He didn't have to volunteer his time and gas to make sure I got to my destination dry and safe, but for some reason he *did*. It was a stunning contrast to a typical Uber ride, such a remarkably heartwarming experience that it made me wonder: *Why don't people always treat one another this way?*

⬭

Later that day after my interview, I wandered into a fashionable modern restaurant called Bartaco in Bucktown, Chicago. I was starved, and three tacos on the menu immediately caught my eye: the glazed pork belly, the Baja fish, and the sesame rib eye.

I like to play a fun game when I eat out. I choose a few items on the menu that I'm interested in and then ask the server to surprise me. Of course I tell them there are no wrong decisions, and I'll be happy with whatever comes out. I mostly do this for the fun (when the food arrives, it's always exciting), but research also shows that upending expectations can improve memory recall, so the added surprise might make the meal more memorable. Plus, I'm indecisive.

I asked my server to surprise me with whichever taco was her favorite of the three. She smiled at me, and I should have known then. A few minutes later, the manager—Scott—came to my table and introduced himself, revealing a platter containing all three tacos. "I know you only asked for one," he said, "but we wanted you to try all three for the same price."

I couldn't believe it. I offered my deepest thanks, and he accepted them genuinely. It was a special moment: two strangers building a bond in a taco shop. After enjoying my delicious tacos (the glazed pork belly was the best . . . see, surprise *does* help you remember things better!), I went up to the bar to talk with the two of them. I learned that Scott had gone to college in my hometown. After a lovely conversation, I sat back down and asked for the check, but they insisted that my meal was on them. "Don't worry about it," Scott said with a smile. "Pay it forward." I simply couldn't believe this. It was ridiculous—how could someone be so kind? Why is everyone so nice in Chicago?

I thought back to my ride with Fernando earlier that day and was filled with a profound sense of gratitude and warmth. It was a euphoric feeling that I've only rarely experienced in my life. I felt cared for, full of love and companionship. I reflected on how fortunate I am to be able to experience these emotions. I felt *lucky* to be a human on planet Earth, like I was chosen to be part of something big and purposeful. Why had I forgotten what that feels like? For the first time in a long while, I was thankful to not exist on this planet alone. I felt *proud* to be human.

⊙⊙

While writing this book, I came to a major conclusion about the meaning of life. I think that love, social connection, and relationships are why we're here.

I know, I know . . . this may sound ridiculous, but hear me out. We humans were placed here by some force—it makes no difference whether that force is evolutionary or religious or something else. The important questions are: What *keeps* us here? What gives us a will to *live*? What kept our ancient ancestors from feeding themselves to nature, and what stops us from lying down and never getting up?

To succeed as a species, we must be motivated to stay here and remain alive. There must be something that makes us *want* to exist and *enjoy* existing, and I believe that "something" is the love we have for one another. Without the joy we experience from our social bonds, would life be worth living? There's nothing more powerful than the love for a child, a partner, or a best friend. For many, the thought of leaving these people behind is too painful to embrace death.

Research supports this idea. Sadly, lonelier people are roughly five times more likely to die by suicide. Many lives have surely been saved by thoughts of loved ones. For a species that often wonders why it exists and what we are here for, it would only make sense to look to the most powerful experiences that we are blessed with. And what is more powerful than love?

The other day, I sat with my eighty-seven-year-old grandmother at her wooden kitchen table for nearly four hours, talking about life. "I can't believe I'm still alive," she told me. "Everyone I know is gone." It was a sobering thought. Both of Bubbie's parents had died before her twenty-seventh birthday, and as the youngest of six kids, she has now outlived each of her brothers and sisters. She's watched every single one of her lifelong friends pass. I couldn't help but think about this idea that our relationships keep us alive, but as Bubbie had grown older, almost all her cherished connections had atrophied and vanished. In the process, had she gradually come to accept and welcome the inevitable end of her own life? Is this why she was surprised to continue waking up every morning? She seemingly took the thoughts out of my head. "I'm still here because of the love I have for you," she said, gesturing toward me and my mom.

In one statement, Bubbie had both confirmed and upended my suspicion. I was right—her long life *was* being sustained by love. However, it hadn't faded with her lost friends and relatives. As the hand of time pressed unrelentingly forward, shoving former generations into extinction, her social brain had found connection with those who remained around her. Each generation of life is but a new bead of wax dripping down the same candle. As my bubbie's bead was nearing the bottom and drizzling to its culmination, she was finding new life in the beads closer to the flame. We were standing on the shoulders of those who came before us, nurturing my bubbie's brain through love. I was thankful I had gone to visit her.

⊚

You could make a convincing argument that this book is about the brain. Or maybe it's about the benefits of socializing—why we should view one another as medicine. Perhaps it's about the evolutionary origins of our social disposition. But the truth is, this book is about humanity.

It's about you, me, our neighbors, friends, and families, and the invisible barriers we've unnecessarily constructed between us. It's about what happens when you tune into the person sitting across from you on the train, the neighbor listening to music on his porch, the couple walking their dog across your lawn, and your dog barking defiantly in response. It's about the cashier who helps you at the grocery store, the agent who checks your bag at the airport, and the person who answers the phone when you call the bank.

It's about what happens when we let the barriers between us melt and drip away—those rare moments when we see others for their best qualities instead of their worst. It's about the emotions that overtake you when you're spellbound by a gripping conversation and you forget to check the time. It's about getting lost in connection and letting your brain rejoice.

To live in loneliness is a great disservice to the biology we've been gifted with. Engaging with the people around you nurtures your biology; it directs its growth and safeguards against its inevitable decline. Even better, it passes those benefits on to your interaction partner. The way that Fernando and Scott treated me had a colossal impact on my mood and perspective. They uplifted me tremendously, and they both probably live happier, healthier lives because of their social habits. What if everyone adopted such an approach to interacting with strangers?

Instead of finding reasons to connect with strangers, we often inject unnecessary distance and anger between us. When you're sitting in traffic and someone veers in front of you or honks their horn at you, it's easy to get angry and flip them off. Or you could give them grace. As kumbaya as it sounds, I try to give people the benefit of the doubt whenever I can. You never know what someone else is going through. Maybe that stranger in traffic just got news that their parent was rushed to the hospital, and they're trying to get there as fast as they can. Life is a lot more tolerable and less hostile when you approach others this way. And as an added benefit, you create less distance between yourself and your community.

I hope this book might change your mind about connection, helping you see it as not just something you should do, but something your brain *needs*. What does that look like? Well, when your phone starts buzzing on your desk, think hard before you press decline. When you make eye contact with the person on the train, offer a compliment and see if you can learn something from them. Instead of complaining that your neighbor is playing the radio too loud on his porch, bring over a few beers and ask about his taste in music. Instead of navigating life with our attention turned inward, focusing only on the monologue in our heads, we should open ourselves to interaction and let others peek in. Thank the cashier at your grocery store, smile at the person who helps check your bag, and be nice to the bank agent who picks up the phone.

We need to stop neglecting and start connecting for the sake of our brains. Meet up with an old friend or have dinner with family. Consider the quality of your social diet and the ingredients that comprise it. Prioritize relationships that bring you joy. Walk through the grocery store and chat with the cashier. Make friends and don't flake on them. Spend more time with the elderly. Play with your dog. Be a good person, and please, don't get into fights on the internet.

This book is about noticing where we stand as a civilization today and deciding where we're heading tomorrow. We can't individually control the winding path of humankind, but we can each control the path of our own lives. As you read this now, whenever *now* is, we're all standing together at some waypoint and squinting ahead anxiously into a misty, uncertain future. It might be a future in which humanity continues to drift apart, slipping into isolation as automated services take over our interactions and division haunts us. Or it may be a future where we come together, where the joy of being united triumphs against the convenience of being apart. I believe this choice is ours. We can let the crashing waves of time and technology bump against us and force us to drift apart, or we can grab one another and hold tight. I hope we will hold tight, because we *need* one another.

When our neighbors feel more like foes than friends and political disagreements burn hot enough to extinguish lifelong friendships, just remember that we need one another. During life's painful stretches when we're more anxious than content, don't forget that there's comfort in companionship. In a post-interaction world, where we can wander invisibly through digital communities with our faces hidden behind phone screens, it can be tempting to relieve your frustrations by lashing out at others. If you ever feel like yelling at someone you don't know, turn to hug someone you love instead. When temptation strikes, remember that a little empathy can go a very long way.

It's critical that we come together. Enjoy life with people you care about. Tell them you love them. Put a smile on their face. Whatever you do, please do not forget that *we need one another*.

Your brain is counting on you.

# SOCIAL JOURNAL

———— ○ ————

**Instructions:** After a social event or interaction, complete this sheet to reflect on your experience. To download a printable version of this sheet, please visit benrein.com/book.

1.  Short description of the interaction: _____

_____

2.  My favorite part was: _____

_____

3.  My least favorite part was: _____

_____

4.  Describe your mood going in: _____

_____

5.  Describe your mood leaving: _____

    _____

6.  Who did you see? _____

    _____

7.  How well do you know this person / these people? _____

    _____

8.  How much you like this person / these people? _____

    _____

9.  Did you enjoy their company? _____

    _____

10. What did you talk about? _____

    _____

11. Favorite conversation topic: _____

    _____

    Least favorite: _____

    _____

12. What time of day was it? _____

    _____

13. Where were you? _____

_____

14. It was: **(A)** busy / **(B)** quiet _____ I would've preferred

it: **(A)** busier / **(B)** less busy / **(C)** neither _____

15. Were you comfortable in the environment? _____

_____

16. What did you like about the setting? _____

_____

17. What did you dislike about the setting? _____

_____

18. How long were you there? _____

_____

19. The interaction felt like it: **(A)** ended too soon /

**(B)** lasted too long / **(C)** was the perfect length _____

20. If you answered A or B, how much longer or shorter do

you wish it was? _____

_____

_____

21.   How much other interaction have I had today? _____

_____

22.   How much interaction have I had this week? _____

_____

23.   This interaction felt: **(A)** energizing / **(B)** draining /

**(C)** neutral / **(D)** other _____

24.   If I could change anything about the experience, it

would be: _____

_____

25.   Next time, I will try to do more of: _____

_____

26.   Next time, I will try to avoid: _____

_____

**Overall experience score** (circle your answer):

1   2   3   4   5   6   7   8   9   10

1 = worst experience possible; would never do this again
10 = best experience possible; wish it was always this way

# ACKNOWLEDGMENTS

———————— ○ ————————

There's a big difference between reading about a topic and living it. As I wrote this book, I reflected with gratitude on all the people who have shown me the value of social connection firsthand. I am immensely appreciative of my wonderful family, friends, and colleagues, and even the many strangers who have shaped my social experience on Earth. *Thank you* for showing me the meaning of life.

I want to first recognize my earliest scientific mentors who ignited my passion for this field: Drs. Dan McNeil, Elizabeth Levelle, Steven Kinsey, George Spirou, Hawley Montgomery-Downs, Andrew Dacks, and the many other people at West Virginia University who nurtured my initial scientific curiosity. To my academic advisor, Darcey Powell, thank you for mapping out my route to a three-year graduation. You showed me that sometimes reaching for the seemingly unattainable can yield surprising results. That lesson has never left me.

I am grateful for Drs. Zhen Yan, Zijun Wang, Jamal Williams, Freddy Zhang, Megan Conrow, Qing Cao, Treefa Shwani, Ping

Zhong, Luye Qin, Max Rapanelli, Piyali Chakraborty, and the many others who guided me through my PhD at SUNY Buffalo. I had no idea how lucky I was to work in your company. Also, a huge thanks to Drs. Fraser Sim, Malcolm Slaughter, Mikhail Pletnikov, Steven Lewis, James Catlin, Will Mangione, and Thomas Covey, along with Elizabeth White, Brittany Sandor, and the entire JSMBS community for their invaluable support.

To Dr. Rob Malenka, thank you for everything you taught me. Working with you, I learned that doing good science means starting with the experiment that scares you the most . . . and how to tactfully use the phrase "with all due respect." I am grateful for Drs. Monique Smith and Boris Heifets for their guidance and supervision, and for Lindsay Cameron, Matt Pomrenze, Kendall Raymond, Daniel Cardozo-Pinto, Jason Tucciarone, Neir Eshel, Robyn St. Laurent, Zoe Zhang, Ting Wu, Jenny Imamura, Epiphani Simmons, Joshua Crapser, Melanie Plastini, Tuuli Hietamies, and the many other members of the Stanford community who helped make my postdoc productive and enjoyable.

I owe a debt of gratitude to Cleo Abram, Rachel Barr, Julia Bauman, Moha Bensofia, Shawn Carbonell, Brian S. Cohen, John Delony, David Eagleman, Lindsay Ejoh, Harris Eyre, Kimberly Fiock, Julie Fratantoni, Susanna Harris, Andrew Huberman, Inna Kanevsky, Chloe Kirk, Cole Kraten, Jim Kwik, Gabrielle Lyon, Andrei Mayer, Kati Morton, Nini Muñoz, Kathy Nickerson, Barry O'Reilly, Michael Pollan, Corey Powell, Karan Rajan, Erik Reis, Mike Todorovic, Forrest Valkai, Lauren Waldman, Pamela Weintraub, and the many other science communicators who have advised, mentored, and inspired me. Thank you for your guidance and your commitment to science communication. You each make a huge difference.

I am indebted to the many organizations that have supported my work, including the National Institutes of Health, the National Academies of Science Engineering and Medicine, the Society for Neuroscience, Schmidt Sciences, the Pulitzer Center on Crisis Reporting, the UCSC Science Communication Program, the American Association for the Advancement of Science, Sigma Xi, the University of Minnesota, the Allen Institute, the Dana Foundation, the UT Dallas Center for Brain Health, SUNY Buffalo, Stanford University, West Virginia University, *OpenMind* magazine, Arrived-AI, Vlogbrothers, Roon, Mendi, *Diversity in Action* magazine, Nobody Studios, and many others. A very special shout-out to Meriam Good and the entire Mind Science Foundation universe!

I am eternally grateful for my talented agent, Lauren Hall, for her invaluable counsel. I couldn't have done this without you. Let's run it back. I am also beholden to my outstanding publishing team, Anna Paustenbach, Elisabeth Koyfman, and the rest of the group at Avery for their hard work and dedication. Your guidance through this process has taught me invaluable lessons. I would also like to thank Nina Shield for her early belief in this book.

A special thanks to those who offered feedback on the early pages as they came together: Mahima Samraik, my parents, and my lifelong friends Ricky, Sarah, Kali, Christian, and Joey.

To my incredible wife, Bella, who sat with me and read this *entire book out loud* (and some sections multiple times) to give feedback, thank you for investing innumerable hours in this project and many others. I will never stop appreciating you for everything you contribute to my life. You are the best. I love you. I would also like to extend my gratitude to the entire Stephan family. I am so blessed to be an honorary member. Thank you for the care, friendship, and love you've offered me.

To my dog, Zoey, thank you for requesting endless walks and games of fetch to air me out during the writing process. I appreciate you keeping life fun and finding ways to distract and entertain me. You are a really good girl.

And of course to my parents and sister, who raised me with empathy and love, always provided a nurturing and safe home, never stopped encouraging me, and always thought I would write a book. Mom, you showed me the value of kindness and believed in me way before I believed in myself. Thank you for always holding steady in that conviction, even when my confidence has waned.

Dad, thank you for always answering my calls and questions about life. I love and cherish you all.

Last, but most important, I am profoundly grateful for everyone who has connected with my work—whether by watching a video, reading this book, or engaging in any other way. And yes, that includes *you* reading this right now; in fact, if you're still reading, you deserve an especially special THANK YOU! In a world vying for our attention with cheap distractions and questionable information, I deeply and genuinely appreciate your donating a slice of your attention to learning about science. I hope that you've found value in learning about the brain. It is a magnificent personal honor to be able to bring neuroscience into your life. May your life be better for it.

# INDEX

———— ○ ————

Note: Italicized page numbers indicate material in illustrations.

# HOME GROWN

GROSSOLOGY

## ICKY THINGS IN
## YOUR EVERYDAY LIFE

BY
### SYLVIA BRANZEI

ILLUSTRATED BY
### JACK KEELY

GROSSET & DUNLAP

GROSSET & DUNLAP
An imprint of Penguin Random House LLC
1745 Broadway, New York, New York 10019

First published in the United States of America by Grosset & Dunlap,
an imprint of Penguin Random House LLC, 2025

Visit us online at penguinrandomhouse.com.

Library of Congress Cataloging-in-Publication Data is available.

Manufactured in China

ISBN 9780593752418                    10 9 8 7 6 5 4 3 2 1 TOPL

Design by Kimberley Sampson